Ant–plant interactions in Australia

Geobotany 4

Series Editor

M. J. A. WERGER

DR W. JUNK PUBLISHERS   THE HAGUE – BOSTON – LONDON 1982

Ant–plant interactions
in Australia

edited by

RALF C. BUCKLEY

DR W. JUNK PUBLISHERS   THE HAGUE – BOSTON – LONDON 1982

Distributors:

for the United States and Canada

Kluwer Boston, Inc.
190 Old Derby Street
Hingham, MA 02043
USA

for all other countries

Kluwer Academic Publishers Group
Distribution Center
P.O. Box 322
3300 AH Dordrecht
The Netherlands

Library of Congress Cataloging in Publication Data
Main entry under title:

Ant-plant interactions in Australia.

(Geobotany; 4)
'A world bibliography of ant-plant interactions [by] R. C. Buckley' – P.
Contents: Seed utilization by harvester ants / E. A. Davison – Rate of decline of some soil
seed populations during a drought in western NSW / M. Westoby, J. M. Cousins &
A. C. Grice – Relationship between the seed-harvesting ants and the plant community in a
semi-arid environment / D. T. Briese – [etc.]

1. Ants–Australia–Ecology. 2. Insect-plant relationships–Australia. 3. Insects–Ecology.
4. Insects–Australia–Ecology. I. Buckley, Ralf. II. Series.
QL568.F7A56   582'.05'264   82-15344
                                          AACR2

39,407

ISBN 90 6193 684-5   (this volume)
ISBN 90 6193 895-3   (series)

Cover design: Max Velthuijs

PRINTED IN THE NETHERLANDS

Preface

Early research on ant–plant interactions in Australia was largely confined to the economically important problem of ants harvesting surface-sown pasture seed (e.g. Campbell 1966). The report by Berg (1975) of widespread myrmecochory in Australia, and a burst of overseas research, stimulated research on a range of ant–plant interactions in Australia. This book summarizes such research and presents recent and current work on seed harvesting, myrmecochory, ant-epiphytes, extrafloral nectaries, ant–plant–homopteran systems, and the influence of vegetation on ant faunas. I hope that it will encourage further work in these and related areas, and that the review and bibliography of ant–plant interactions in the rest of the world will serve as a useful source for those entering the field. The richness of Australia's flora and ant fauna render it a particularly interesting continent for the study of interactions between them. As immediately apparent from the list of contents, ant–seed interactions are particularly significant in Australia. This is not surprising for a relatively dry continent bearing a largely sclerophyllous plant cover. Future research, however, especially in the tropical north, is likely to reveal further types of interaction, perhaps corresponding to those characteristic of the tropics elsewhere, or perhaps distinctively Australian.

Some of the chapters have been shortened and modified considerably from the original manuscripts, but the ideas and results presented are, of course, those of the individual authors. To those who were particularly prompt in preparation of their original manuscripts, notably Mark Westoby and Camilla Huxley, I owe an apology for the time spent in assembling the rest of the book. One compensation: the hypothesis that the concentration of myrmecochory in Australian sclerophyll vegetation is due largely to soil nutrient deficiencies, advanced by Westoby et al. in Chapter 8, receives support from the comparative studies of Milewski and Bond, described in Chapter 9.

Besides the individual authors themselves, I should like to thank Marcia Murphy, Jenny Tode and Caroline Twang for typing and retyping the manuscripts and revisions.

Contents

Principal authors' addresses

Andersen, Alan. School of Botany, Melbourne University, Parkville, Victoria 3052, Australia.

Briese. David T., Division of Entomology, C.S.I.R.O., P.O. Box 1700, Canberra, A.C.T. 2600, Australia.

Buckley, Ralf C., Department of Biogeography and Geomorphology, Research School of Pacific Studies, Australian National University, P.O. Box 4, Canberra, A.C.T. 2601, Australia; present address: AMDEL, P.O. Box 114, Eastwood, SA 5063, Australia.

Campbell, Malcolm H., Agricultural Research Centre, Forest Road, Orange, N.S.W, 2800, Australia.

Davison, Elizabeth A. Department of Zoology, University of New England, Armidale, N.S.W. 2351, Australia.

Fox, Marilyn D. Royal Botanic Gardens and National Herbarium, Sydney, N.S.W. 2000, Australia.

Huxley, Camilla R. Department of Botany, Oxford University, South Parks Road, Oxford OX1 3RA, England.

Majer, Jonathan D. Department of Biology, West Australian Institute of Technology, Hayman Road, Bentley, W. A. 6102, Australia.

Milewski, Anthony V., Western Australian Herbarium, Department of Agriculture, George Street, South Perth, W. A. 6151, Australia.

Westoby, Mark. School of Biological Sciences, Macquarie University, North Ryde, N.S.W. 2113, Australia.

CHAPTER ONE

Seed utilization by harvester ants

Elizabeth A. Davison

Abstract. Two species of harvester ants, *Chelaner whitei* and *C. rothsteini*, were studied in the far west of N.S.W. Data collected in both field and laboratory were used to study seed utilization and the effect of seed supply upon harvester ant populations. Nest density, foraging activity and colony reproduction were monitored throughout the study. Four nests of each species were excavated (one in each season) to determine nest structure, colony size and seasonal changes in colony composition. There were 156 colonies of *C. whitei* and 66 colonies of *C. rothsteini* on the study plot. Colonies ranged in size from 450 to 40,000 workers *(C. whitei);* and from 15,000 to 58,000 *(C. rothsteini). C. whitei* foraged all year round. *C. rothsteini* ceased foraging during winter. Peak foraging by both species occurred in summer, when both species predominantly gathered seeds of ephemeral grasses and forbs. Differences in seed selection by the two species were related to size, phenology and chemical composition of the seeds and fruit. All seeds taken into the colonies were eventually fed to the larvae. Food digested by the larvae was transmitted through the colony by trophallaxis (from larvae to the workers). The final stage larvae of *C. rothsteini* were found to act as a storage caste resulting in decreased foraging activity by this species when sufficient larval fat stores had accumulated. It is postulated that seed supply is the most important factor regulating populations of harvester ants. Finally, speculations are advanced concerning possible reciprocal long-term evolutionary effects of the interaction between harvester ants and plants.

Introduction

In arid and semi-arid habitats around the world, many species of insects, birds and small mammals rely on seeds as their main source of energy input. Granivory is especially advantageous to animals living in unpredictable, xeric habitats, since many desert plants produce large quantities of nutrient-rich seeds which can resist desiccation and survive in a dormant condition for many years. During drought, starvation may produce drastic reductions in the populations of herbivores dependent on fresh plant foods, and of predators dependent on such herbivores. Populations of seed-storing granivores, however, are assured of a reliable food supply long after the vegetation has dried up, and are therefore more stable. In the arid regions of Australia, seed-harvesting ants are the most important granivores, and the

characteristic crater-shaped middens of discarded seed husks and soil are a conspicuous feature of the Australian arid landscape.

In 1975 a four-year study of two species of harvester ants, *Chelaner whitei* and *C. rothsteini,* was undertaken at Fowlers Gap Arid Zone Research Station in the far west of New South Wales. These two were the most abundant of the nine species of harvester ants found on the alluvial plains at Fowlers Gap. Data were collected in both field and laboratory to assess the type and amount of seed removed by these ants, the utilization of seed within the colony and the effect of seed supply upon the ant populations. Nest density, foraging activity, and colony reproduction of the two species of *Chelaner* were monitored throughout the study, on a 1.5 ha plot situated in a moderately grazed paddock on the alluvial plains. Four nests of each species were excavated, one each in spring,

Buckley, R. C. (ed.), Ant-plant interactions in Australia.
© 1982, Dr W. Junk Publishers, The Hague. ISBN 90 6193 684 5.

1

summer, autumn and winter, to determine nest structure, colony size and seasonal changes in colony composition.

Results and discussion

Colony density of both species remained almost constant throughout the study, with 156 colonies of *C. whitei* per ha and 66 of *C. rothsteini*. The excavated nests of *C. whitei* ranged from 1 to almost 6 m depth and those of *C. rothsteini* from 1 to 2 m. Individual colony populations ranged from ca 450–40,000 adult workers for *C. whitei* and from ca. 15,000–58,000 adult workers for *C. rothsteini*.

C. whitei foraged all year round, whereas *C. rothsteini* ceased foraging during winter. Peak foraging activity of both species occurred in summer, following favourable spring rains. At such time, both species predominately gathered seeds of ephemeral grasses and forbs. Samples of forage material taken from returning foragers, and contents of nest granary chambers, were collected and identified to assess seasonal changes in the composition of forage. Results are shown in Fig. 1 and Table 1: *C. whitei* foragers collected seeds from 20 species of plants, representing 14 genera; *C. rothsteini* collected seeds from 10 species of plants in 9 genera. The diet of *C. rothsteini* consisted mostly of grass seeds (27–92%), *Plantago* seeds (3–37%) and composite seeds (up to 32%). *C. whitei* took smaller proportions of grasses (3 to 10%) and more forb and chenopod seeds; up to 65% composite seeds and up to 80% chenopod seeds. There are seasonal trends in the diets of both species, reflecting differences in seed availability: both took a greater proportion of ephemeral forb and grass seeds when these were abundant after favourable rains in winter 1975, summer 1975/76 and winter 1978. During these periods, foraging activity was intense and both species were observed removing seeds directly from the plants of ephemeral species such as *Tripogon loliiformis, Eragrostis dielsii, Dactyloctenium radulans, Brachycome* spp., and *Minuria cunninghamii*, for *C. rothsteini*, and *Tripogon loliiformis, Dactyloctenium radulans, Brachycome marginata,* and *Helipterum floribundum*, for *C. whitei*. As conditions became progressively drier during 1976, 1977 and the first half of 1978, the seeds of many ephemeral forbs and grasses comprised progressively smaller proportions of both ants' diets, perhaps because such

Table 1. Granary contents of *Chelaner whitei* nest excavations. From Davison (1980).

Species	Number of seeds	Weight (g)
Nest excavated: June 1976		
Atriplex angulata (large)	1168	0.461
Atriplex angulata (small)	813	0.698
Bassia spp.	834	0.690
Chloris truncata	1835	0.329
Dactyloctenium radulans	6442	2.208
Dichanthium sericeum	1251	0.584
Enneapogon avenaceus	396	0.142
Enneapogon cylindricus	1543	0.484
Panicum decompositum	1501	1.042
Portulaca oleracea	5629	0.751
Sida spp.	104	0.100
Nest excavated: May 1977		
Astrebla lappacea	36	0.046
Atriplex angulata	66	0.060
Atriplex spp.	450	0.389
Bassia spp. (small)	36	0.022
Bassia spp. (large)	78	0.072
Dactyloctenium radulans	36	0.012
Enneapogon avenaceus	174	0.061
Nest excavated: October 1977		
Dactyloctenium radulans	13450	4.573
Enneapogon avenaceus	100	0.035
Enneapogon cylindricus	8100	2.301
Unknown grass 2	50	0.028
Nest excavated: December 1977		
Astrebla lappacea	9	0.011
Atriplex angulata	235	0.282
Atriplex spp.	2	0.002
Bassia spp. (small)	1419	1.277
Bassia spp. (large)	136	0.164
Dactyloctenium radulans	263	0.092
Dichanthium sericeum	600	0.390
Enneapogon avenaceus	12750	4.516
Enneapogon cylindricus	399	0.113
Panicum decompositum	902	0.574
Unknown grass 1	50	0.051
Unknown grass 2	46	0.027

seeds were harder to find after dispersal. Under these conditions, foragers collected more seed from abundantly-seeding species whose seeds were less susceptible to wind dispersal and thus remained concentrated under the plants: e.g. *Dactyloctenium radulans, Plantago varia, Atriplex* spp., *Bassia* spp., and *Boerhavia diffusa*. Seeds of the ephemeral species, which were easily dispersed by wind, were generally smaller (0.037 to 0.49 mg) than those which remained concentrated under the plant (0.35 to 1.74 mg).

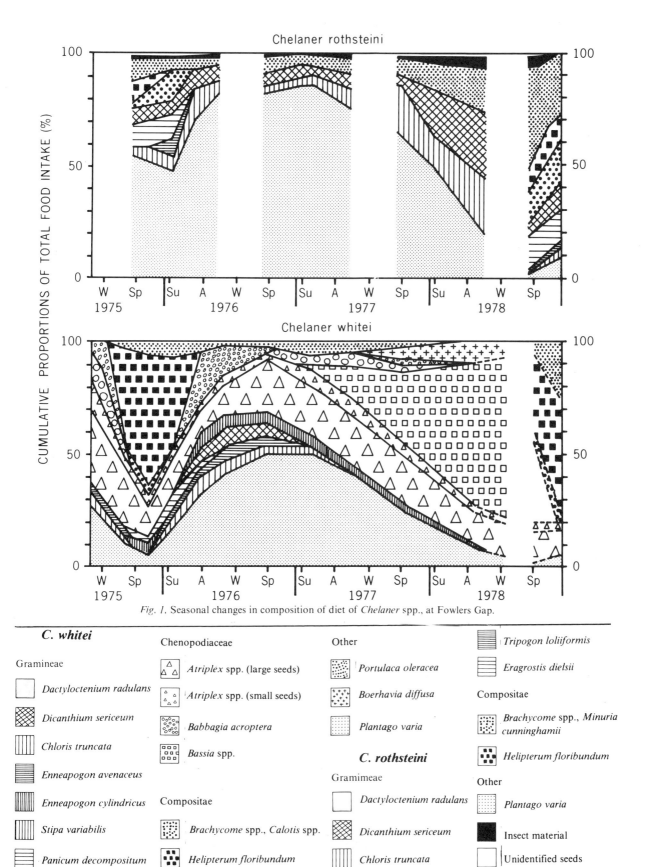

Fig. 1. Seasonal changes in composition of diet of *Chelaner* spp., at Fowlers Gap.

C. whitei

Gramineae

☐	*Dactyloctenium radulans*
⊠	*Dicanthium sericeum*
⫿	*Chloris truncata*
▤	*Enneapogon avenaceus*
▥	*Enneapogon cylindricus*
⫿	*Stipa variabilis*
▤	*Panicum decompositum*

Chenopodiaceae

△	*Atriplex* spp. (large seeds)
△	*Atriplex* spp. (small seeds)
◎	*Babbagia acroptera*
▦	*Bassia* spp.

Compositae

▨	*Brachycome* spp., *Calotis* spp.
▦	*Helipterum floribundum*

Other

▨	*Portulaca oleracea*
✚	*Boerhavia diffusa*
▨	*Plantago varia*

C. rothsteini

Gramineae

☐	*Dactyloctenium radulans*
⊠	*Dicanthium sericeum*
⫿	*Chloris truncata*

▤	*Tripogon loliiformis*
▤	*Eragrostis dielsii*

Compositae

▨	*Brachycome* spp., *Minuria cunninghamii*
▦	*Helipterum floribundum*

Other

▨	*Plantago varia*
■	Insect material
☐	Unidentified seeds

The annual seed biomass per unit area removed by these two species was determined from seasonal studies of foraging activity: the mean daily number of foraging colonies, the mean number of successful foraging trips per colony per day, and seasonal changes in forage composition (Tables 2 and 3). No effective rainfall was recorded at Fowlers Gap between winter 1976 and autumn 1978. The drought resulted in a general reduction in foraging, overriding the seasonal oscillations in foraging activity. The annual consumption of both species fell by about half from 1976 to 1977: from 11.57 to 4.85 kcal/m^2 for *C. whitei* and from 13.63 to 7.22 kcal/m^2 for *C. rothsteini*.

Differences in seed selection by the two ant species are related to the size, phenology and chemical composition of seeds and fruit. Both species collected seeds in the same overall size range, but the smaller *C. rothsteini* foragers (2.5 to 3.0 mm body length) collected proportionally more seeds in the smaller size classes than the larger *C. whitei* foragers (3.0 to 6.5 mm body length). *C. rothsteini* collected very few chenopod seeds. The seeds of most chenopods at Fowlers Gap are encased in a large spongy bractiole (up to 10 mm diameter in some *Atriplex* spp.) or in a large woody pericarp with long spines (up to 4 mm diameter in some *Bassia* spp.), which makes these seeds very difficult for the small *C. rothsteini* foragers to carry back to the nest. On the other hand, *C. rothsteini* foragers collected the large *Plantago* seeds which are usually encountered free of any encumbering plant material. The larger *C. whitei* foragers collected seeds from 0.33 to 1.74 mg in weight, and fruit up to 15 mg. They were able to carry several seeds at once, as in a whole grass spikelet or a whole composite flower head. The smaller *C. whitei* foragers, in contrast, collected only smaller seeds: 0.08 to 0.45 mg. Seed or fruit shape may affect handling

Table 2. Seasonal estimates of caloric content of food collected by *Chelaner whitei* foragers, per unit area. Based on caloric contents per gram. From Kendeigh and West (1965).

Season/year	Gramineae (kcal/m^2)	Chenopodiaceae (kcal/m^2)	Compositae (kcal/m^2)	*Plantago* (kcal/m^2)	Other (kcal/m^2)	Total (kcal/m^2)
Spring 1975	0.255	0.829	1.300	0.367	–	2.751
Summer 1975–1976	0.982	2.334	3.374	1.316	–	8.006
Autumn 1976	0.844	0.726	–	0.255	–	1.855
Winter 1976	0.157	0.158	–	0.040	–	0.355
Spring 1976	0.513	0.918	–	0.035	0.030	1.496
Summer 1976–1977	1.720	3.239	–	0.514	0.055	5.528
Autumn 1977	0.531	3.088	–	0.311	0.062	3.993
Winter 1977	0.096	0.340	–	0.100	0.014	0.550
Spring 1977	0.127	1.250	–	0.124	0.140	1.641
Summer 1977–1978	0.689	1.367	–	0.024	0.030	2.110
Autumn 1978	0.028	0.651	–	–	0.017	0.676
Winter 1978	0.006	0.416	–	–	0.005	0.423

Table 3. Seasonal estimates of caloric content of food collected by *Chelaner rothsteini* foragers, per unit area.

Season/year	Gramineae (kcal/m^2)	Compositae (kcal/m^2)	*Plantago* (kcal/m^2)	Insect* (kcal/m^2)	Total
Spring 1976	1.694	–	0.866	0.020	2.580
Summer 1976–1977	7.750	–	2.634	0.044	10.248
Autumn 1977	0.670	–	0.128	0.008	0.806
Winter 1977	–	–	–	–	–
Spring 1977	0.824	–	0.080	0.020	0.924
Summer 1977–1978	3.174	–	2.156	0.170	5.500
Autumn 1978	0.448	–	0.324	0.024	0.796
Winter 1978	–	–	–	–	–

* Assuming mean value of 6.0 kcal/g for insect material.

efficiency: this may account for the lack of certain grasses such as *Panicum decompositum* and *Enneapogon* spp. from the diet of *C. rothsteini*. These species have smooth round seeds, carried only with difficulty by *C. rothsteini* foragers, but more readily by *C. whitei* foragers, which have better developed clypeal teeth.

Seasonal changes in ant forage composition at Fowlers Gap are due to differences in plant phenology. Ephemeral grass seeds are most abundant following effective rainfall in spring or early summer, and peak foraging activity of both species of *Chelaner* coincides with such periods. Rainfall from 2 months beforehand was more highly correlated with foraging activity than rain which fell earlier or later. Many ephemeral forb seeds, however, are produced with effective rains during the cooler months: such seeds are generally too large to be harvested by *C. rothsteini*, but are collected by *C. whitei*. Hence the latter can forage all year round, whilst the former is inactive aboveground during winter: each species is most active when its preferred seeds are most abundant.

Observations on marked *C. whitei* foragers show that individuals return repeatedly to the same spot to collect the same seed type, suggesting that foragers of this species may form a chemical search image for particular types of seeds. Detailed chemical analyses

Table 4. Protein content and weight of seeds.

Seed type	Weight per seed (mg)	% Protein content
Gramineae		
Panicum decompositum	0.637	26.92
Unidentified	0.582	4.31
Chloris truncata	0.217	23.22
Enneapogon avenaceus	0.354	23.16
Enneapogon cylindricus	0.284	20.34
Stipa variabilis	0.488	18.56
Dactyloctenium radulans	0.351	18.20
Tripogon loliiformis	0.076	–
Eragrostis dielsii	0.037	–
Chenopodiaceae		
Bassia ventricosa	0.920	33.21
Bassia diacantha	0.601	29.79
Atriplex angulata	0.894	27.29
Atriplex vesicaria	0.794	25.77
Atriplex lindleyi	1.743	25.33
Plantago varia	1.350	16.34
Portulaca oleracea	0.133	–

and behavioural experiments to test for differences in seed attractiveness were not made, but the protein contents of various seeds are shown in Table 4. Chenopods generally had higher protein content than grass seeds: 25.3–33.2% as opposed to 18.2–26.9%. Differences in seed protein contents may account for the presence or absence of insect or other animal matter in the natural diet of harvester ants. *C. whitei* were apparently able to satisfy their protein requirements with seeds alone, since chenopod seeds have a higher protein content than grass seeds: *C. rothsteini*, however, were unable to collect enough of the chenopod seeds, owing to the size and weight of the pericarp, and supplemented their diet with animal matter to make up for the protein deficit. Harvester ants may also select seeds preferentially on the basis of oil content or total energy content: oils yield more calories per unit weight than carbohydrates or protein (Southwood 1966). The mean oil content of grasses (5.18%) is less than that of chenopods (8.86%), and that of composites (24.89%) greatly exceeds both (Levin 1974): the same general pattern can be assumed for seeds at Fowler's Gap. Preference for oil-rich seeds may account for the high proportion of the composite *Helipterum floribundum* in the diet of *C. whitei* in the spring of 1975 and 1978, when ephemeral grass and chenopod seeds were also abundant. Carroll and Janzen (1973) suggested that the presence or absence of toxins may also be an important factor in seed selection by harvester ants. The fact that grass seeds are generally low in toxins (Janzen 1971) may contribute to the high proportions of grass seeds in the diets of both species of *Chelaner* at Fowlers Gap. Overall it appears that within the constraints of seed or fruit size and shape the chemical factors outlined above also influence seed selection by these two *Chelaner* species.

The seeds (husked or unhusked) are carried into the nest by returning foragers and stored in underground granary chambers. Laboratory observations revealed that food is transmitted through colonies of the two *Chelaner* species as follows: (a) the workers collect and prepare (de-husk, cut up or chew) the seeds for presentation to the larvae, (b) larva digest the seeds, (c) adults imbibe nutrient-rich stomodeal (oral) secretions of the larvae, and (d) workers regurgitate liquid to other workers and reproductives by trophallaxis, thereby distributing nutrients throughout the colony. No workers, of either species, were ever observed to feed directly on the seeds: the adult ants of these species appear to be totally dependent upon the larvae for their immediate nutrition. In addition, proctodeal or anal trophallaxis was also observed in *C. rothsteini* colonies: workers intensively licked and imbibed anal secretions of the large final-stage larvae. These larvae may function as a storage caste since they were found to contain large amounts of fat, suggesting that much of the starch in the grass seeds which comprised the bulk of their diet was converted to fat for storage.

Larvae were found in nests of both species in all seasons. The greatest number were found in spring and summer, coinciding with peak foraging activity for both species. *C. whitei* foraged all year round, but *C. rothsteini* were totally inactive above ground during winter, with little or no seed stored in underground granaries. Since *C. rothsteini* did not forage to feed the overwintering larvae, the latter must have been able to sustain themselves, the adult workers and queen with the fat they had accumulated during summer and autumn. This contention was supported statistically by the finding of a strong negative correlation between foraging activity and larval biomass in *C. rothsteini* but not in *C. whitei*.

Whilst temperature and humidity determine whether foraging occurs, food supply and demand regulate foraging intensity. Food supply depends primarily on recent rainfall and temperature. Food demand depends upon the numbers and degree of satiation of larvae within the colony. Food supply regulates demand in that fewer larvae survive to adulthood when food is scarce, and the number of reproductives is greatly reduced, whilst the reverse is true when food is plentiful.

Seed supply thus appears to be the most important factor regulating populations of these harvester ants. Are there any reciprocal effects of the ants on the vegetation? Ants at Fowlers Gap may contribute to plant diversity by selectively harvesting the more abundant seeds, but it is equally possible that selective harvesting of rare species (e.g. *Tripogon loliiformis*) could render them even rarer. On a longer time scale, a number of the preferred forage plants may have evolved mechanisms to reduce depredation of their seed stocks by harvester ants; notably the longer-lived shrubs whose seeds are encased in hard woody pericarps covered with hairs or long spines (e.g. *Bassia* spp., *Salsola kali*) or with large spongy bractioles *(Atriplex* spp.).

References

Carroll, C. R. & Janzen, D. H., 1973. Ecology of foraging by ants. Ann. Rev. Ecol. Syst. 4: 231–257.

Davison, E. A., 1980. Ecological studies of two species of harvester ants, *Chelaner whitei* and *C. rothsteini,* in an arid habitat in south eastern Australia (Hymenoptera: Formicidae). Ph. D. Thesis, U.N.E., Armidale.

Janzen, D. H., 1971. Seed predation by animals. Ann. Rev. Ecol. Syst. 2: 465–492.

Kendeigh, S. C. & West, G. C., 1965. Caloric values of plant seeds eaten by birds. Ecology 46: 553–555.

Levin, D. A., 1974. The oil content of seeds: an ecological perspective. Am. Nat. 108: 193–206.

Southwood, T. R. E., 1966. Ecological Methods with Particular Reference to the Study of Insect Populations. Methuen, London.

CHAPTER TWO

Rate of decline of some soil seed populations during drought in western New South Wales

Mark Westoby, J. M. Cousins and A. C. Grice

Abstract. Soil seed populations of three common grass species did not decline detectably over 16 months of drought at Fowlers Gap, N.S.W., despite continued harvester ant activity. These data contrast with evidence from North America showing that desert soil seed reserves are substantially depleted both by ants and by rodents.

Introduction

We have collected data on soil seed reserves at a desert research site over a period of drought which followed heavy rains and consequent heavy seed-set. Here we report aspects relevant to the effects of predation on seed populations.

Methods

This work was done at Fowlers Gap Research Station, in western New South Wales. Average rainfall is about 200 mm per year, with almost no seasonality (Fig. 1). Chenopod shrublands are the main vegetation types in the area; in some cases the perennial shrubs have been overgrazed and the dominants are now short-lived shrubs and grasses. The general

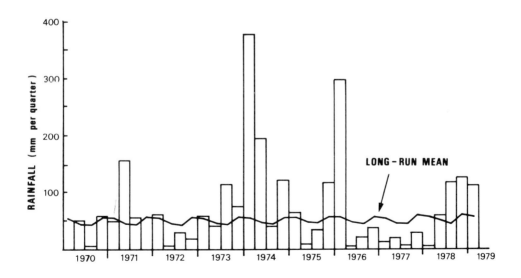

Fig. 1. Rainfall totals (mm) at Fowlers Gap Station in quarter-years from July 1970 to March 1979.

Buckley, R. C. (ed.), Ant-plant interactions in Australia.
© 1982, Dr W. Junk Publishers, The Hague. ISBN 90 6193 684 5.

ecology of the station is described in Mabbutt (1973).

We extracted seeds from 482 fifty-gram samples of soil. These were collected from a large range of locations in January 1976, August 1976, January 1977, August 1977, November 1977, and December 1977. One to three replicate samples were taken from each location. Most locations were sampled repeatedly, but the set sampled at each date was not completely the same.

Samples were collected as follows. Five soil blocks, each 20×20 cm $\times 4$ cm deep, were taken at 2 m spacing along a transect. These were mixed, and a 500 g sample taken from the aggregate. The contents of stones and of moisture were determined for each 500 g sample, so that a given weight of air-dry, stone-free soil could be regarded as representing the top 4 cm of a known area of soil surface. These samples were then split into fifty-gram subsamples. This aggregation and subsampling reduced the variance between samples, compared to the variance there would have been between 50 g samples taken directly from the field soil. Variance was still high, however.

One 50 g split from each 500 g sample was mixed into concentrated KCl solution with specific gravity 1.55. Organic matter floated and was removed, rinsed and dried. The seeds in this material were identified and counted under a binocular microscope. Other 50 g subsamples were spread thinly on sterilized soil and kept moist under various conditions of temperature and day length, and the germinating seedlings counted until emergence stopped. Correction coefficients were obtained for the flotation and counting procedures, to allow absolute density to be estimated. These will not be described here, since only relative changes over time are discussed.

This study was preceded by moderate winter rains and heavy summer rains (Fig. 1); both led to substantial inputs of seed to the soil reserves. There were no further rains producing growth and seed-set until May 1978, after the end of this study. It was therefore possible to examine the decline in the soil seed reserves of individual species in the absence of further input. We did not attempt to calculate rates of decline, (a) for species at locations where their mean seed density was less than 5 000 per m^2, or about 5 seeds per 50 g sample, or (b) for species which released their seed slowly, so that there was continuing input to the soil during the drought, or (c) where there were reasons to believe the extraction procedure was unreliable, as in *Crassula* and *Wahlenbergia*. These exclusions left three grass species, *Dactyloctenium radulans*, *Sporobolus actinocladus*, and *Enneapogon avenaceus*. For locations where the mean estimated density of one of these species was greater than 5 000/m^2, log density of that species was regressed against time for the last five dates only. This corresponds to an exponential decay model: there was no evidence that any other model would have fitted the results better, nor that locations with lower mean densities behaved differently.

Results

The regressions characterizing the exponential decline rate varied greatly. The density of each species appeared to rise at some locations, decline at others, and remain approximately constant at others. There were no tendencies to any more complex pattern (e.g. an increase followed by a decrease), when all sites were considered together, and variances were too large to allow such patterns to be detected at individual sites. The mean regression slopes were not significantly different from zero for any of the species (Table 1): any decline in the soil seed population of

Table 1. Rate of decline of soil seed densities with time: means and standard errors of regression slopes for regressions of log (density/ m^2 + 1) on time in years for selected species and locations (see text).

Species	Number of regressions	Mean slope	Equivalent percent survivorship per annum	Mean ± s.d. of density/ m^2 at locations included in regressions
Dactyloctenium radulans	25	−0.067 ± 0.077	86	26,000 ± 12,000
Sporobolus actinocladus	15	−0.093 ± 0.125	81	50,000 ± 44,000
Enneapogon avenaceus	6	−0.025 ± 0.188	94	7,700 ± 1,700

these three species was too small to be detectable.

This conclusion is consistent with the data of Davison (this volume) which imply that the harvester ants *Chelaner rothsteini* and *C. whitei* removed 5,000–10,000 *Dactyloctenium* seeds per m² over two years from her study area. *Dactyloctenium* was one of the main species harvested. *Dactyloctenium* densities at the locations from which our regressions came ranged from 10,000/m² to 50,000/m², with a mean of 26,000/m² (Table 1). Taking these data together, a decline in the range 20–35% over the two years of the drought might be expected for a typical *Dactyloctenium* population, which is close to what we observed (Table 1).

Discussion

Many desert plant species maintain large reserves of soil seed, and granivory is therefore an important way of life for desert consumers. Considerable indirect evidence, recently reviewed by Brown, Reichman and Davidson (1979), indicates that desert granivores compete with each other. This evidence implies that some granivore species are capable of significantly reducing the supply of seeds for other species. More direct evidence that desert soil seed reserves are substantially affected by predation comes entirely from North America. Various calculations have suggested that consumption is a large proportion of seed production (Tevis 1958; Chew and Chew 1970; Soholt 1973; French et al. 1974; Nelson and Chew 1977; Pulliam and Brand 1975; Whitford 1978). Recently exclusion experiments with direct measurement of densities of soil seeds larger than 0.4 mg (Brown et al. 1979) have shown predation effects that were consistent in different treatments, although often not statistically significant in any one comparison.

How can our data, showing no decline in particular seed populations despite continued ant activity, be reconciled with this evidence? The possibilities are as follows.

Firstly, it should be recognized that the failure to detect a decline is partly due to the high variance of the data: but such high variance is found in all soil seed data, and cannot be used to reject our observations but accept others. In fact, the confidence intervals on our data are relatively small, owing to the large number of samples (e.g. about 250 for the

Dactyloctenium radulans regressions as a group) and the mechanical averaging over 5 quadrats of 20×20 cm within each sample. Thus for *Dactyloctenium*, even if the true mean rate of decline were two standard errors greater than our estimate, it would only be about 40% per year, which does not suggest that seed predators are limited by the numbers of seeds present.

Secondly, predation rate could have been small relative to seed density because predation rate was particularly low during this particular time period, or seed densities particularly high. Lacking data from any other time periods, we can not exclude these possibilities. The first seems unlikely; harvester ants remained active throughout the drought (Davison, this volume). The drought was preceded by three summers of above-average rainfall (Fig. 1), of which 1973–4 and 1975–6 were exceptionally wet. It may be that seed densities were unusually high.

A third possibility is that predators may compete for and consume a significant proportion of the seeds of particular preferred species only. This would not be reflected in detectable decline rates in less-preferred species. However, *Dactyloctenium* is an important food of harvester ants at Fowlers Gap (Davison, this volume). Brown, Davidson and Reichman (1979) excluded some small-seeded species from analysis because of their extreme between-sample variance and because they were thought not to be an important resource for granivores. However, the density of *Filago californica*, which has seeds of similar size (4×10^{-5} g) to *Sporobolus actinocladus*, the smallest-seeded of the species in Table 1, was influenced by predation in North America.

Finally, it is possible that predation pressure on seeds may be lower in Australia than in North America, or alternatively lower at Fowlers Gap than at other sites studied. Australian deserts have a higher probability of extended droughts, for a given mean annual rainfall, than do deserts on other continents (Low 1978). It has been argued (Morton 1979) that this might make life difficult for the few rodent granivores in Australia, but it is hard to see why it should limit seed-harvesting by ants. The most likely mechanism limiting harvesting is that seeds might be buried faster in the cracking clay loams of Fowlers Gap than in the sandy or gravelly soils studied to date in North America, thus becoming unavailable to ants. This is suggested by the circumstance that ant-foraging activity decreased markedly in the second year of the drought (Davison, this volume), despite the high seed densities that were still present.

Evidence is not yet available to distinguish all these possibilities. On the whole, the most likely seems to be that Fowlers Gap is different from sandy sites, owing perhaps to different seed burial rates. In addition, the depletion rate of seeds may have appeared relatively small because of an unusually high density at the start of the drought. In any event, it is premature to assume that all arid zone seed populations are under substantial predation pressure at all times. This must be reckoned with when discussing competition between different granivore species.

Acknowledgments

This work was funded by the Australian Department of the Environment. We thank the staff of Fowlers Gap Research Station for much help and hospitality.

References

Brown, J. H., Davidson, D. W. & Reichman, O. J., 1979. An experimental study of competition between seed-eating desert rodents and ants. Am. Zool. 19: 1129–1144.

Brown, J. H., Reichman, O. J. & Davidson, D. W., 1979. Granivory in desert ecosystems. Ann. Rev. Ecol. Syst. 10: 201–227.

Chew, R. M. & Chew, A. E., 1970. Energy relationships of the mammals of a desert shrub *(Larrea tridentata)* community. Ecol. Monogr. 40: 1–21.

French, N. R., Maza, B. G., Hill, H. O., Aschwanden, A. P. & Kaaz, H. W., 1974. A population study of irradiated desert rodents. Ecol. Monogr. 44: 45–72.

Low, B. S., 1978. Environmental uncertainty and the parental strategies of marsupials and placentals. Am. Nat. 112: 197–213.

Mabbutt, J. A. (ed.), 1973. Lands of Fowlers Gap Station, N.S.W. University of N.S.W. Research Series, 3.

Morton, S. R., 1979. Diversity of desert-dwelling mammals: a comparison of Australia and North America. J. Mammal. 60: 253–264.

Nelson, J. F. & Chew, R. M., 1977. Factors affecting the seed reserves in the soil of a Mojave Desert ecosystem, Rock Valley, Nye County, Nevada. Am. Midl. Nat. 97: 300–320.

Pulliam, H. R. & Brand, M. R., 1975. The production and utilization of seeds in plains grasslands of southeastern Arizona. Ecology 56: 1158–1166.

Soholt, L. F., 1973. Consumption of primary production by a population of kangaroo rats *(Dipodomys merriami)* in the Mojave desert. Ecol. Monogr. 43: 357–376.

Tevis, L., 1958. Interrelations between the harvester ant *Veromessor pergandei* (Mayr.) and some desert ephemerals. Ecology 39: 695–704.

Whitford, W. G., 1978. Foraging in seed-harvester ants *Pogonomyrmex* spp. Ecology 59: 185–189.

CHAPTER THREE

Relationship between the seed-harvesting ants and the plant community in a semi-arid environment

D. T. Briese

Abstract. The five species of seed-harvesting ants at a study site in semi-arid New South Wales comprise the most abundant trophic group of the ant community. These ants and the plant community are described briefly, and areas of potential interaction, which could influence the population dynamics of either community, are defined. A study of these interactions suggests that there are no strong short-term regulatory effects, and that the granivore–plant system is rather loosely coupled. Plant population dynamics are determined primarily by the timing and magnitude of an unpredictable rainfall, with little feed-back effect from seed-harvester activity. The ants do not appear to be limited by seed availability, and a system of buffers within the ant community dampens population fluctuations when food resources change. Such loose coupling within and between trophic levels would tend to stabilize the structure of communities subjected to extreme, variable and unpredictable environments, such as the semi-arid and arid regions of Australia.

Introduction

Ants in Australia are ubiquitous, abundant and highly active, making them one of the most important animal groups in terms of energy flow (Brown & Taylor 1970). The semi-arid regions of the continent support a very rich ant fauna (Greenslade 1979; Morton, in press), and, in a study of a semi-arid saltbush area, Briese (1974) found that seed-harvesting species were the most abundant class of ants. An increase in the abundance and importance of seed-harvesting ants as habitats increase in aridity has been noted generally (Sudd 1967; Stradling 1978) and, in Australia, they constitute the major granivores of the arid regions (Morton 1979). The relationship between these seed-harvesting ants and the plants from which they derive their food would seem to be one of the more important in the overall functioning of semi-arid and arid environments. The aim of this chapter is to examine the community of seed-harvesting ants in a semi-arid habitat from the viewpoint of their interactions with the plant community.

There are two questions basic to the subject of ant–plant interactions, involving the effect of vegetation on the structure and dynamics of associated ant populations, and conversely the effect of the ants on the structure and dynamics of the plant populations supporting them. Other aspects such as possible co-evolution must be considered in the light of such information. Where and when do these effects occur?

Harper & White (1970) have proposed a model for the population dynamics of plants which incorporate four stages at which regulatory effects might take place: (1) the seed bank – the accumulated reserve of seeds present in and on the soil; (2) an environmental 'sieve' due to the heterogeneity of the soil environment and other vegetation, whose various factors determine whether or not a seed will germinate; (3) changes in plant number and size as seedling populations develop; (4) production of new seeds which will contribute to the next seed-rain.

A similar 4-stage model may also be used to describe ant population dynamics: (1) the release of sexuals, and dispersal of newly-mated queens; (2) an environmental 'sieve' which determines survival of newly-mated queens, and which would be due prim-

Buckley, R. C. (ed.), Ant-plant interactions in Australia.
© 1982, Dr W. Junk Publishers, The Hague. ISBN 90 6193 684 5.

arily to predation, intra specific competition or the location of favourable nest sites (Briese 1974); (3) changes in colony numbers and size as they grow and develop; (4) production of new sexuals, which will contribute to subsequent generations.

A simple diagram of these two models, indicating the areas where potential interactions could take place, is shown in Fig. 1. In the case of seed-harvesting ants, the major interactions with plants would occur at stages P1 and P2. Ants may reduce the seed-rain by removing unripe propagules from plants, or can reduce the accumulated reserve of seed by collecting them from the ground surface. As soil-inhabiting animals, they could also subtly influence stage P2, by modifying the factors in the soil environment which act as a sieve *sensu* Harper and White.

The main areas of interaction by plants upon ants would occur in stages A3 and A4. The structural diversity of plants could influence nest site availability and favourableness, but more importantly the quality and quantity of food provided by the plant species will determine the survival, rates of growth and reproductive capabilities of seed-harvester ant colonies.

Following the outlines of these models, this chapter will first describe the elements of the interacting plant and harvester ant communities. Second, it will examine the influence of these seed harvesters on the seed bank, to determine whether they might modify the absolute or relative abundance of the plants, or

alter their spatial distribution. Next, it will examine the ways in which plant structure and seed availability might influence the species composition and spatial distribution of harvester ant colonies.

Potential areas of interaction between plant and harvester ant communities do exist, and these could lead to feed-back mechanisms and co-evolution of the communities. In the light of this analysis, it should be possible to determine the nature and extent of the ant–plant interactions occurring at one particular site, and to offer an insight into the general nature of such relationships.

The plant community

The semi-arid pasture areas of the Riverine Plain in southern New South Wales can basically be divided into two categories – saltbush and grassland. The grassland is a degraded community that has developed under heavy grazing from a saltbush-dominated climax community (Moore 1953).

The climate of the region is characterized by its high variability. The mean annual rainfall at the study site is 412 mm, but wide and erratic fluctuations occur (range 145–812 mm), and there is no consistent seasonal incidence of rainfall. High evaporative losses reduce the effectiveness of summer and autumn rainfalls. Temperature also varies widely; the mean monthly maximum and minimum range

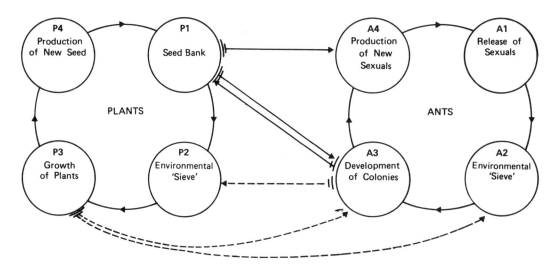

Fig. 1. Stages in the population dynamics of a plant community and a harvester ant community, showing sites of potential interaction (see text for details).

from 31 °C and 15 °C in February to 13 °C and 3 °C in July. Extremes of 47 °C and –2 °C were recorded during the study period. The number and severity of winter frosts is also highly variable.

The site chosen for the study was just inside the southern limit of the saltbush community, at 'Emmet Vale' (latitude 35°06'S, longitude 144°48'W, elevation 86 m), a property 51 km north of Deniliquin. It has been described in detail by Briese & Macauley (1977); it is flat, with uniform grey clay soil and structurally simple vegetation. This facilitated the study by reducing the number of factors which might affect both the plant and ant communities. The site contained a saltbush area subjected to light sheep grazing, and a grassland area in which prior heavy grazing had caused marked changes in plant composition (Wilson, Leigh & Mulham 1969). Observations were made between 1971 and 1973.

The natural vegetation was a shrub-steppe formation dominated by bladder saltbush *Atriplex vesicaria* Hev. ex. Benth., with cotton bush *Maireana aphylla* R.Br., and soft-horned saltbush *Malacocera tricornis* (Benth.) R. H. Anders, occurring as scattered plants. The area between these long-lived perennial shrubs was occupied by a ground flora of numerous annual and perennial grasses and forbs. These varied markedly in abundance, depending on rainfall. The most common were the grasses *Danthonia caespitosa* Gaudich, *Sporobolus caroli* Mez, and *Chloris truncata* R.Br.; the dwarf chenopods *Maireana pentagona* R. H. Anders and *Maireana cheeli* R. H. Anders; and the forbs *Medicago polymorpha* L., *Minuria cunninghamii* (DC.) Benth. and *Brachycome* spp.

In the heavily-grazed area the shrub layer had been eliminated, and replaced by a grassland, in a state of successional change. The perennial grass *Danthonia* was the most stable dominant plant, while favourable rainfalls led periodically to marked increases in the semi-annual grasses *Chloris* and *Sporobolus*, and they then became the most abundant plant species. The populations of the dwarf chenopods *Maireana* spp. were also considerably greater in the grassland than in the saltbush.

With regard to their interaction with seed-harvesting ants, the most important part of the plant species' life cycles is the flowering period. On this basis the plant community can be divided into two groups; those which flower only once per year, usually in spring as a response to increasing day length (Willi-

ams 1961), and those which normally flower in spring, but also at other times in response to effective rainfalls. The flowering patterns of plant species contributing more than 2% of seeds to the diet of harvesting ants in the saltbush or grass areas are shown in Fig. 2. The patterns of three species included in this group – *Chloris truncata*, *Sporobolus caroli* and *Minuria cunninghamii* – are complicated because they grow either as short-lived perennials in favourable seasons, or as annuals when conditions are unfavourable. Consequently, they did not flower following the dry winter and spring of 1972, but responded to heavy summer and autumn rains in 1973.

The incidence and amount of rain are not only important to flowering, but also strongly influence seedling populations of species such as *Danthonia* (Williams 1968). In a study of a *Danthonia*-dominated grassland in this region, Williams (1961) found he could arrange the plant species in a sequence from those showing predominantly cool-season growth, through those which could grow all year, to those which grew predominantly in the warm seasons. However, the occurrence of effective rains again dictated the periods available for growth, especially in the dominant perennial species. In fact, an abnormal-

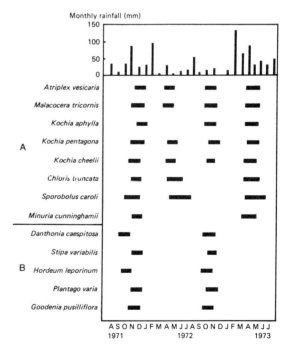

Fig. 2. Periods of seed production for the more common plant species at the study site. A. Plants flowering more than once per year. B. Plants flowering only once per year.

13

ly high rainfall can substantially alter the composition of the ground flora in both saltbush and in grassland areas (Williams 1968).

Generally, the life cycles of the species making up the plant communities in this semi-arid region are geared to the wide climatic variability they experience, particularly the unpredictable rainfall (Williams 1961), and it is in this regard that their interactions with the ant community must be considered.

The harvester ants

The composition of the ant community at the study site has been described in detail by Briese & Macauley (1977). Over the two-year study period 37 species were collected, and these could be divided into categories based on their food usage (Table 1). Of the 37 species, the majority were highly omnivorous, and only 10 collected seeds. Of these, 5 species were classed as seed-collecting omnivores, since seeds comprised only a minor part of their diet. Seed or plant material formed over 90% of the diet in the remaining 5 species. They were classed as seed harvesters. It is this group of ants, and their interactions with the plant community, upon which this paper will concentrate.

The species are *Chelaner* sp. A (group of *C. whitei* (Wheeler)), *Chelaner* sp.B. (group of *C. rothsteini* (Forel)), and *Pheidole* sp. A, B and C. This labelling is used to maintain consistency with previously published articles on the ant community (Briese & Macauley 1977, 1980, 1981; Briese 1982a). Appropriately labelled voucher specimens have been placed in the Australian National Insect Collection at the C.S.I.R.O. Division of Entomology, Canberra, *Pheidole* is a cosmopolitan genus, while *Chelaner* is endemic to Australian region. Both genera are well known for seed-harvesting species (Brown & Taylor 1970; Greenslade 1979), but little has been published

Table 1. Functional categories of the ant species.

Category	Species[a]	Number of species	Percentage of total colonies	
			Saltbush	Grass
Specialist predator	*Cerapachys* sp. *Sphinctomyrmex* sp.	2	unknown	
Generalist predator	*Odontomachus* sp.	1	1	7
Predator--scavenger	*Rhytidoponera* spp. (3) *Melophorus* sp.	4	11	14
Omnivore (nectar collecting)	*Iridomyrmex* spp. (5) *Camponotus* spp. (4) *Polyrhachis* sp.	10	31	14
Omnivore (seed collecting)	*Melophorus* sp. *Xiphomyrmex* sp. *Meranoplus* spp. (3)	5	12	10
Seed harvester	*Chelaner* spp. (2) *Pheidole* spp. (3)	5	42	53
Unknown	*Crematogaster* sp. *Epopostruma* sp. *Mesostruma* sp. *Monomorium* sp. *Oligomyrmex* sp. *Tapinoma* spp. (2) *Camponotus* sp. *Iridomyrmex* sp. *Melophorus* sp.	10	3	2

[a] Bracketed values are the number of species per genus.

14

on this aspect of their behaviour.

Chelaner sp. A has a relatively large black head, reddish thorax and orange gaster; the workers are polymorphic, ranging in size from 3 mm to 6 mm. Colony size ranges up to 1,000–2,000 workers. The ants forage in trails, with larger workers predominating, and seed is carried back and stored in the nest galleries. The nests are conspicuous because of large surface mounds (up to 30 cm in diameter) of compacted seed husk and earth around their entrances. *Chelaner* sp.B has a reddish-brown head and thorax, and black gaster. There is no worker polymorphism, all individuals being about 2.5 mm long. The colony size is relatively large (to over 50,000 workers) and the ants forage in long trails (up to 40 m); seed is collected and stored in nest galleries; pebbles are also collected and used to decorate the nest mound, which may have one or more entrances. *Pheidole* sp.A was the largest of the three *Pheidole* species present. It is a black ant with two distinct worker castes; the minors are about 2 mm long, and the majors about 5 mm long, with enlarged reddish heads. Foraging is carried out almost entirely by the minors, on short trails (up to 5 m) to seed patches. Individual seeds are brought back and stored in the nest galleries. The major workers function chiefly as defenders of the colony, which may contain up to ca. 10,000 worker ants. The nests have one to four entrances commonly surrounded by small cones of discarded seed-husk and chaff. *Pheidole* sp.B was the smallest member of this genus present. It has similar caste structure to *Pheidole* sp.A, with black minor workers about 1.2 mm long, and all-black majors about 3 mm long. Foraging activities are also similar, but colony sizes are smaller, with less than 1,000 workers. The nests rarely have seed-husk material discarded around their single entrances. *Pheidole* sp.C was intermediate in size to the other two species, the minor workers being brown-black and about 1.6 mm long, while the majors reach 4 mm in length. In all other respects this species is very similar to *Pheidole* sp.A, though its colony sizes are probably slightly smaller.

Colonies of these five ants were distributed throughout the study plots (Fig. 3) and, while activity patterns of the species varied (Fig. 4), all five were active at times when the major seed crops were produced (Briese & Macauley 1981). Even though this group comprised only 5 of the 37 species found in the region, they formed 42% of the total number of colonies in the saltbush and 53% in the grassland area

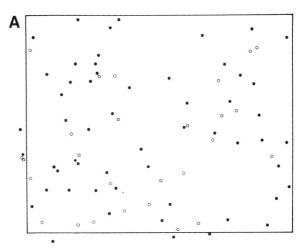

Fig. 3. Distribution of colonies of harvester ants. A. The saltbush site. B. The grassland site. *Pheidole* sp.A ●; *Pheidole* sp.B ■; *Pheidole* sp.C ○; *Chelaner* sp.A □; *Chelaner* sp.B *.

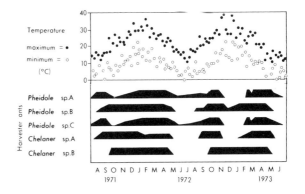

Fig. 4. Seasonal activity patterns of the harvester ants.

15

(Table 1). This implies that the seed harvesters form an important part of the overall ant–plant relationships in this particular semi-arid environment. The interactions between these two communities will now be examined more closely.

Do ants affect the plant community?

As mentioned in the introduction, the influences of harvester ants on the plant community may occur in two broad areas; through the seedbank, because of their granivorous habit, or by modification of the soil environment because of their nesting behaviour. The more obvious and direct interaction is via the seed bank, and through it the ants can affect the plant community in three ways: they can reduce the absolute abundance of vegetation, change its relative abundance due to selective foraging, or alter the spatial distribution of plant species by transporting seeds from one area to another. Modification of the physical and chemical properties of the soil by ants may also contribute to this last effect (Briese 1982b).

Absolute abundance of plants

The effect of seed harvesting on annuals and short-lived perennials, whose population dynamics depend to a large extent on the establishment and growth of successive generations, is quite different from their effect on long-lived perennials such as *Atriplex vesicaria,* which has an estimated life span of 16 to 24 years (Hall et al. 1964). In stable populations of such species, reproduction by seedlings is not very important for population maintenance (Harper & White 1970). The survival of mature bushes is the critical factor, since there is normally a chance for the survival of new seedlings only upon the demise of an older plant. Although there are large seedling recruitments, few of these generally persist to maturity. The possibility of a reduction in absolute abundance of plants in each of these categories will therefore be considered separately.

With regard to annual and short-lived perennial species, field observations following seasonal rains revealed very high seedling populations in both the saltbush and grassland areas, with ground flora estimates of 90 and 154 plants/m² respectively (Briese 1974). This probably underestimates the total germination, for the populations were subsequently re-

duced through self-thinning, interplant competition and climatic stress to densities of 14–16 and 46–65 plants/m² respectively, at a later stage or plant maturity. This suggests that factors affecting seedling survival (the third stage of Harper & White's model) would exert the most critical influences on plant populations.

The high levels of germination observed indicate that there is a very large reserve of seeds of the annual species in and on the soil. Subsequent self-thinning of the plants showed that the germination rates did not limit plant populations, and therefore that possible ant modifications of the seed bank could not have substantially affected final population size. Briese (1974) estimated that the ants collected less than 20% of the available crop of freshly fallen seed (over 20,000 seeds/m² produced in the saltbush area during an intense period of harvesting in autumn of 1973 (Table 2). As there were no other major seed predators, most of the seed crop therefore reaches the seed bank. The ants prefer freshly fallen seed (Briese & Macauley 1981), so that even if they were to harvest on occasion the bulk of a year's seed crop, the seed bank would protect the annual plant community from a subsequent reduction in absolute abundance.

Since they rely less on new seedlings to maintain the population, long-lived perennials would appear even more 'resistant' to seed-harvesting activities, and perhaps would be expected to tolerate a greater seed loss. Measurements taken during the same period, however, comparing the removal of seed from beneath exposed bushes with that from bushes protected by ant exclosures, indicated a low rate of seed

Table 2. Estimated amount of seed taken in a 500 m² saltbush plot by harvester ants during autumn 1973.

Species	Seed taken (g) [a]
Pheidole sp.A	110
Chelaner sp.B	60
Chelaner sp.A	35
Pheidole sp.C	10
Pheidole sp.B	1
Total taken	216
Total produced [b]	1200
Percentage of seed production harvested by ants	18%

[a] Based on foraging rates and total time foraged.
[b] Estimated from measurements of plant density and seed production per plant.

removal, averaging 24% of the current crop (Briese 1974). This is comparable to that for annuals. There was, however, wide variability between test bushes (0–53%), probably a result of the ant species' habit of foraging intensively in particular patches (Briese & Macauley 1981). Such short-term patchiness in harvesting was also observed by Williams (1972), who found that ants had removed from 0 to 86% of seeds that had fallen under individual bushes of *Atriplex vesicaria* in other parts of the Riverine Plain. Thus, with the dominant perennial, as with the ground flora, the harvesting ants had negligible impact on the absolute abundance of plants.

Relative abundance of plants

The different ant species collected different seeds (Table 3). Hence the ants might affect the relative abundance of plant species through selective foraging.

The proportions of each seed in each ant diet may be greater or less than the proportion of that seed in the overall seed-rain. Thus, for example, *Chelaner* sp.A concentrated on *Atriplex* in the saltbush plot, while the smaller *Pheidole* sp.B. took mainly small grass seeds.

The proportions of seeds collected by the ant community as a whole, however, more closely reflected the proportions observed in the available seed pool. There was no obvious bias toward seeds of any of the rarer plant species. This indicates that the ants did not contribute to the rareness of particular plants by selective harvesting, as was observed by Tevis (1958) in the case of the North American seed-harvester ants. Any possible bias was rather toward the selection of seeds from the dominant plant species such as *Atriplex, Sporobolus, Chloris* and *Danthonia*. This occurred because the ants were attracted to heavy concentrations of freshly fallen seed (Briese & Macauley 1981). The combined effect of the seed harves-

Table 3. Forage items taken by harvester ants over a 2-year period in the saltbush and grassland plots as percentage of total number.

Forage item[a]	*Chelaner* sp. A		*Chelaner* sp. B		*Pheidole* sp. A		*Pheidole* sp. B		*Pheidole* sp. C	
	saltbush	grassland	saltbush	grassland	saltbush	grassland	saltbush	grassland	saltbush	grassland
Atriplex vesicaria	72.6		12.7		38.0		1		6	
Malococera tricornis	7.2		0.8		1.8					
Maireana aphylla	1.0		0.2		0.7					
M. pentagona	4.3	26.0		0.2	0.3	9.0			1	
M. cheelii	0.7	9.4	0.3	0.2	0.2	4.3				
Bassia brachyptera	1.2	0.2	0.6		0.6	0.6				
Danthonia caespitosa	0.9	17.2	4.5	8.5	1.0	10.5		5	1	
Sporobolus caroli	1.2	0.8	29.2	36.8	19.4	39.7	68	66	46	
Chloris truncata	1.3	18.7	15.9	25.9	0.5	3.6	5	3	5	
Vulpia myuros	0.8		0.4	0.4	0.1	0.5				
Stipa variabilis		11.0	0.1	1.5		0.1				
Hordeum leporinum		2.9	0.8			0.6				
Plantago varia	0.8	1.0	3.3	0.6	0.2	4.3			2	
Minuria cunninghami		0.2	10.6	10.7				1		
Medicago polymorpha	0.8	0.2	1.5	1.2	1.8	1.7	2		2	
Goodenia pusilliflora	0.5	4.5		1.1		4.2				
Brachycome campylocarpa		0.8		0.5						
Calotis hispidula		1.6		0.2						
Daucus glochidiatus		0.2				0.2				
Hypoxis pusilla			0.2	1.2	0.4	0.3	1	1		
Swainsona oriboides			0.3		0.1		1			
Unidentified seeds	1.8	1.7	1.1	1.4	1.1	5.5	2	3		
Other plant matter	4.0	3.3	13.3	7.9	30.0	10.9	13	18	33	
Insect matter	0.4		3.7	2.1	3.6	4.0	7	3	4	
Total number of seeds	564	717	733	663	1130	548	66	73	76	

[a] Forage items are seeds unless otherwise mentioned.

The values are rounded for *Pheidole* spp. B and C because of small sample size.

ters may then be to help maintain plant species diversity by taking proportionately more of the common seeds, rather than to reduce diversity, as would happen were rarer seeds selected. Harper (1969) has suggested that such apostatic selection improves the competitive ability of rarer species, and, in a recent study in North America, Brown et al. (1979) observed a decrease in the diversity of winter annuals following the removal of seed-harvesting ants.

Spatial distribution of plants

As the harvester ants collect seed from one area and move it to another (the nest) it is possible that they might act as seed dispersal agents. This would apply primarily in the case of the larger shrubs, the dwarf chenopods and the large-seeded forbs. The seeds of the common grasses are more subject to wind dispersal, whereas the larger and heavier seeds of the shrubs accumulate beneath the bushes. Williams (1972) has shown that 83% of germination in *A. vesicaria* occurred beneath the female bush. Such plants would have poorer chances of survival than ones which germinated away from mature bushes, so that ants could influence plant pattern by moving the seed about. Moreover, the areas on and around the harvester ant colonies are more favourable sites for plant growth, since the deposition of chaff and plant material by ants resulted in increased levels of carbon, nitrogen and phosphorus in the soil at the nest site (Briese 1982b). Despite this, only three *A. vesicaria* seedlings were observed growing out of nest sites, probably because the ants chew off the radicles of most germinating seeds (Briese 1974).

The densities of dwarf *Maireana* spp. on and around nests of *Chelaner* sp.A was more than double that on control plots. They formed a major part of the seed diet of this species, and the harvesting activity appeared to result in localized dense patches. However, the area involved is only a very small proportion of the total surface area (< 1%), and as these *Maireana* spp. are not uncommon elsewhere the effect is probably quite small.

Do plants affect the ant community?

The study site at 'Emmet Vale' contained an area of degraded grassland, surrounded by the original saltbush plant community. This provided an ideal situa-tion to determine whether plant structure and composition per se could affect the ant community. The grassland had originally been covered with saltbush, and was small enough that a separate fauna could not develop, for the proportions of colonizing ant propagules (e.g. newly-mated queens, ant groups resulting from nest fission) arriving at both areas would be the same. This meant that any differences in the ant community would be due primarily to the differences between the two areas induced by their respective plant communities.

Briese (1974) mapped out the species composition (Table 4) and nest distribution (Fig. 3) of the harvester ants in two 500 m² quadrats in the saltbush and grassland areas. All five species of ant were present in each quadrat, and the total numbers of colonies were very similar. However, there were some differences in the composition of the species (Table 4), mainly a reduction in the number of *Pheidole* sp.A and an increase in the number of *Chelaner* sp.B in the grassland compared to the saltbush. Differences in ant faunas associated with different vegetation types have been well documented (Hayashida 1960; Gregg 1963; Schumacher & Whitford 1976). These are due primarily to differences in available nest sites, foraging substrates and consequently food sources. In the present study, it is unlikely that the observed differences could be due to differences in vegetation structure, since all the ant species were ground-nesting and the spatial distribution of colonies was not influenced by the vegetation (Briese & Macauley 1977). Furthermore, all species of harvester ant foraged almost exclusively on the ground surface, so that the absence of a shrub layer in the grassland would not have affected the foraging substrate. This leaves changes in plant composition, and the consequent changes in available seed, as a possible cause for differences in the ant species composition.

Table 4. The relative abundance of harvester ant colonies in the saltbush and grass plots.

Species	Saltbush	Grass
Pheidole sp. A	42	26
Pheidole sp. B	8	11
Pheidole sp. C	19	26
Chelaner sp. A	9	9
Chelaner sp. B	2	14
Total colonies	80	86

It seems plausible that the increases in nest density of *Chelaner* sp.B, and *Pheidole* spp.B and C in the grassland might have been due to the increased abundance of the smaller grass and forb seeds which they preferentially harvested. However, Table 1, apart from indicating differences in seed collected by different species, also demonstrates that there were substantial differences in the seed taken by a particular species in either the grassland or the saltbush area. The various species showed a very wide dietary range, and demonstrated foraging opportunism; when a seed preferred in one plant community was absent or rare in the other, the ants readily switched to alternative seed species. This plasticity of diet coupled with the relatively small proportion of total seed production actually harvested, reduces the possibility that differences in plant species composition were acting as a determinant of ant species composition in this case.

The behaviour of the ant species suggests that interspecific ant competition contributed to the differences observed. The lower densities of *Pheidole* sp.A in the grassland were associated with higher densities of *Pheidole* sp.C, with the total number of colonies of both species together remaining approximately constant; 52 and 61 colonies in grassland and saltbush respectively. These species are very similar morphologically, overlap considerably in their behaviour, and exhibit interference competition through interspecific hostility (Briese & Macauley 1977). The species react to each other as they would to conspecifics, and therefore it is the total number of colonies of the two species which is important. A third area examined by Briese (1974) also contained a similar total of 55 colonies (37 of sp.A and 18 of sp.C). Briese & Macauley (1977) found that the distribution of nests of these species was regular, suggesting that space was limiting the colony density. It would seem then that the maximum combined density of both species was one colony per 8–10 m². The relative contribution of each species to this depended on prior events relating to interspecific and intraspecific competition, and any differences which might occur between areas are not greatly influenced by plant composition or structure.

There were two large colonies and twelve small colonies of *Chelaner* sp.B in the grassland, compared with only two large colonies in the saltbush. Colonies of this species, however, reproduced by nest fission (Briese 1974); a single large colony can give rise to at least 14 smaller ones. The many small colonies in the grassland probably originated in this way, so that the original number of large colonies may have only been three or four, which does not differ substantially from that of the saltbush area. The biomass of *Chelaner* sp.B supported by each site during the sampling period would have been similar.

Given the above, compositional differences in the ant faunas of the grassland and saltbush are only small. Both areas supported a similar total number of colonies and similar biomass of harvester ants (Table 4). However, the grassland was far more productive because of the greater density of heavy cropping annual and perennial grasses, having an estimated seed production of 75,000 seeds/m² (6.6 g/m²) compared with 35,000/m² (3.3 g/m²) in the saltbush during one period of flowering (Briese 1974). It therefore appears that some common factor limits the overall population density of the harvester ants in both areas to a level below that which would be set by food availability. Differences in productivity, as determined by changes in plant species composition at a particular site, do not appear to have much influence on the harvester ant community.

Community structure and ant–plant interactions

The evidence presented in the previous sections suggests that at this particular site the seed-harvesting ants have had very little effect on the structure, abundance or species composition of the plant community. Likewise, the vegetation has had only a small influence on the structure, composition and colony density of the ant community. The loose coupling of these two communities means that the feed-back links shown in Fig. 1 are very weak, and consequently that co-evolutionary changes would not have been a significant force in the development of the communities. This at first may seem incongruous in a situation where one group of organisms is dependent on another for its food supply, and in the process reduces the reproductive capacity of the other, but is more understandable if we consider the nature of semi-arid and arid ecosystems.

Noy-Meir (1973) considers such ecosystems to have three main attributes: precipitation so low that water becomes the controlling factor for biological processes; a highly variable water input; and a high degree of unpredictability in the timing and magnitude of this input. Consequently, the adaptations of organisms in these ecosystems, especially the primary producers, are geared to survival in an area where favourable periods are relatively short and unpredictable. Many of the plants at the study site exhibited such adaptations; high seed yields and seed longevity results in the build-up of large reserves, while seed polymorphism and complex germination mechanisms ensure that germination is optimized at the most favourable times. Subsequent strong competition for the available groundwater amongst seedlings and mature plants can influence the structure of the vegetation. Production in such a system tends to be of the 'pulse-reserve' type (Noy-Meir 1973), where rain might initiate a pulse of production, some of which is diverted to a reserve (seeds) which only depletes slowly. This acts as a buffer against long unfavourable periods, and forms a base from which the next growth pulse is initiated. The chief structuring forces acting on arid or semi-arid vegetation are the physical limitations, exerted primarily by the low, variable and unpredictable input of water. Subsequent biological pressures, such as those which may be exerted by the harvesting activity of ants, act only in a very minor way.

The adaptive strategies of the harvester ants to these extreme and unpredictable physical conditions fell into two groups. *Pheidole* spp. A and C, and *Chelaner* sp. A generally avoided unfavourable conditions by remaining in their nests for long periods, while *Pheidole* sp.B and *Chelaner* sp.B were much more tolerant of extremes of temperature and aridity (Briese & Macauley 1980). However, all the harvesting ants were very well adapted to take advantage of the 'pulse-reserve' responses of plants. In the present situation, their strategy was to maximize foraging activity during the periods when pulses of seed production occurred (Briese & Macauley 1981). The group-foraging shown by these species was also well adapted to collecting seed which occurred in dense patches, and hence they rapidly accumulated large stores of seed in their underground granaries. These could then act as a buffer against periods of reduced food availability. Colonies from the group of species, mentioned above, which avoided physical extremes were also able to cease activities over periods when conditions were still favourable (Briese & Macauley 1980). In this case, the storage of food led to reduced energy expenditure and risk of losses due to predation etc. The range of seeds taken and the dietary

overlap between ant species varies markedly in response to environmental change (Briese 1982a). When food 'pulses' occurred, different species of ant tended to converge on common food items, but exploitative competition for these was not evident. During periods between pulses when seed became scarcer, a greater range of seeds were taken by each ant, although this often led to greater overlap between specific pairs of ants. This indicates that some weak exploitative competition could occur at low food densities. However, the presence of a large seed bank, and the observation that seed removal was much lower than seed production, suggests that such situations would not occur very frequently. Colony density remained relatively stable over the two and a half years of observations, and data on colony density and spacing indicated that intraspecific and, to some extent, interspecific competition stabilized populations below the level at which food would become limiting (Briese & Macauley 1977). Similarities in colony density between two areas with markedly different levels of seed production support this. An unknown proportion of total harvester ant production was channelled off through higher trophic levels, but most of this was in the form of dead worker ants collected by scavengers (Briese 1974). Few active predators were observed, suggesting that the links between ant populations and their predators were also loose.

The response of the harvester ant community to its environment can be characterized by a tiered buffering system (Table 5), which helps to stabilize community structure in an unpredictably fluctuating environment. Even though the system seems to be held together quite loosely, there are mechanisms which could potentially reduce exploitative competition between the harvester ants, such as differences in activity periods (Briese & Macauley 1980) and differences in type and size of preferred seeds (Briese & Macauley 1981), as well as the self-spacing mechanisms which limit colony density. This suggests that the mechanisms structuring the community are geared to the extremes of the climatic variability, and hence are operating on a time scale much longer than could be observed in the course of this study. Damping of fluctuations through loose coupling would ensure the long-term stability of the system, for if species and trophic groups are too closely linked the unpredictability and magnitude of changes in the system would greatly increase the probability of extinction.

Table 5. Buffering mechanisms which would stabilize ant community structure in an unpredictable environment.

Mechanism	Level	Mode of action
Colony organization	Colony	Dampens fluctuations in ant populations, as changes can be absorbed within colonies without affecting colony number
Food storing	Species – seed harvesters	Protects against periods of low food supply
Aggressive interactions (self-spacing mechanisms)	Species – all ants	Lead to division of space, and limitation of colony densities at levels below those at which food might become limiting; therefore protects against fluctuations in food supply
Loose-coupling between ecologically similar species	Community	Minimizes the possiblity of exploitative competition between species; enables several species to utilize a particular resource class.
Loose-coupling between trophic levels	Community	Reduces the changes of food limitation during periods of low productivity

A diagram of this model is shown in Fig. 5.

Unfortunately, very few comparative data exist for Australia. Some quantitative studies have been made in more temperate areas, indicating that seed-harvesting ants can have considerable impact on the composition and yield of pasture crops following oversowing with exotic seeds (e.g. Campbell 1966; McGowan 1969). Such agricultural systems are not comparable, for the 'natural' plant community is artificially manipulated to produce higher yields of a few selected exotic species, and the impact of the ants can be gauged only in this light. Moreover, the rates of sowing are much less than the seed falls which occur naturally, and the effect of ants is therefore exaggerated under these conditions. In a natural arid plant community in western New South Wales, Westoby et al. (this volume) found that the soil seed population of certain grasses did not decline detecta-

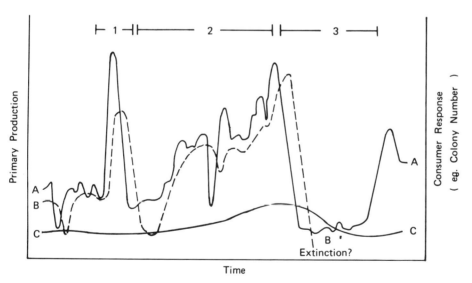

Fig. 5. Hypothetical examples of close-coupling and loose-coupling strategies between a producer and consumer in an unpredictable, fluctuating environment. A = producer (fluctuations reflect climatic variability); B = close-coupled consumer (maximizes own productivity); C = loose-coupled consumer (optimizes own productivity).

Features of loose-coupled consumer: (1) does not respond prematurely to increases of producer; (2) response may increase slowly over long periods of high favourability – overall fluctuations are greatly dampened, and held well below potential levels at a particular time by the internal buffering mechanisms (see text for details); (3) actual resource limitation may occur during prolonged troughs in levels of producer – the response may decrease during this stage.

Features of close-coupled consumer: (1) and (2) response closely follows that of producer at all times. Resource limitation occurs at most times; (3) catastrophic change in producer may lead to overshooting of consumer response, severe resource depletion and possible extinction.

bly over 16 months of drought, despite continued harvester ant activity.

Several studies have recently been published concerning harvester ants (chiefly of the genus *Pogonomyrmex*) in semi-arid and desert areas of North America. It appears that there is much tighter coupling between trophic levels in North America, for Brown et al. (1979) found that the size of the seed bank was significantly increased by the exclusion of ants over a two-year period. This led to an increase in the overall density of winter annuals, although plant biomass remained unaffected. Consequently, the ant community might compete for, and be limited by, food resources at most times. Davidson (1977) and Hansen (1978) have described mechanisms, such as separation of ant species by size, and consequently by size of seeds utilized, and differences in foraging behaviour, which could reduce exploitative competition between species. In the study described here, all ant species showed the same group foraging behaviour, while separation of species by size does not seem to be as important in Australia as in North America (Morton 1982). This may indicate that the granivorous ant–plant system in Australia is more loosely coupled than its American counterpart. It should be noted that comparisons are complicated by the presence of granivorous rodents in North America, which compete strongly for seed resources with the ant species (Brown et al. 1979). Such rodents are relatively unimportant in Australia (Morton 1982). However, the within-habitat diversity of seed-harvesting ants is the same in both regions (Morton, in press) which could imply that there has been no compensation for the absence of granivorous rodents in Australia, but rather a loosening of linkages within the granivore–plant system.

Other results from North America tend to be more compatible with the present findings. Tevis (1958), Rogers (1974) and Whitford (1978a) have all reported that harvester ants took only small proportions of seed production. Preferential harvesting, however, was thought to have some effect on the relative abundance of plant species (Tevis 1958; Whitford 1978b). Davidson (1977) compared ant faunas of several different sites. She found only a poor correlation between the presence of seed-harvesting ants and the diversity of perennial vegetation, but a strong relationship between ant species diversity and productivity, as well as a greater colony density in the more productive sites. This does not necessarily conflict with the finding in the present study that differences in productivity between sites did not affect species composition or density, since the sites here were much more similar, with identical substrate, rainfall and ant faunas. The differences in productivity here may well reflect the effect of short-term fluctuations within a particular region. The loose-coupling hypothesis would argue that the ant communities of both sites were subject to common selection pressures, dictated by long-term variations in productivity. Davidson's estimates of productivity are based on annual mean precipitation, which probably correlates with long-term productivity levels, and would reflect differences in the lower extremes of productivity between sites. Assuming that the system is loosely coupled it is at these productivity troughs that selection pressures would occur. Davidson's data would suggest that the mechanisms are flexible enough to allow species diversity to be optimized, rather than maximized, at different levels under different conditions of aridity. Bernstein & Gobbel (1979) found that overdispersion of colonies, resulting in the limitation of population density, was common in a series of ant communities she studied on an environmental gradient in the Mojave Desert. Moreover, their data led to the postulate that the foraging areas of these colonies were negatively correlated with food density. This indicates one way by which the buffering mechanism might accommodate different conditions of aridity. Bernstein also commented on the stability of the ant communities over the six-year study period, in line with the findings reported here.

Further evidence for the loose-coupling hypothesis can be found in Whitford's (1978b) observation that the impact of harvester ants on seed reserves is a function of the past climatic history. Increased levels of colony satiation lead to reduced harvesting, independent of the amount of seed available. This shows that such a system not only buffers the ant community against extreme fluctuations, but can also protect the plant community by allowing the build-up of seed reserves under more favourable conditions.

In a variable environment there are two strategies that an organism could follow. The first involves a fine-tuning to the environment, in which the population would closely follow changes in productivity of the trophic level beneath it, and so maximize its own productivity at all times. This would require very complex mechanisms for monitoring environmental change, for the danger always exists that, should changes be too severe or occur unexpectedly, the population will overshoot, resulting in depletion of resources and possible extinction. The alternative strategy is a more cautious approach. Population numbers would be held at levels determined more by the lower extremes of long-term environmental fluctuations. This would be carried out by a system of buffers (Table 5) which prevent a premature response and limit the extent of responses to changes in the environment. Consequently, for most of the time population numbers of the organism are much lower than the system could at that particular time support. This loose-coupling strategy would result in a more stable population (Fig. 5).

Harvester ants are particularly suited to such a strategy because of their colonial social structure and use of a storeable food source. Such loose-coupling within and between trophic levels may well be a general property of communities subject to the extreme, variable and unpredictable conditions of Australia's semi-arid and arid regions.

References

Bernstein, R. A. & Gobbel, M., 1979. Partitioning of space in communities of ants. Ecology 48: 931–942.

Briese, D. T., 1974. Ecological Studies on an Ant Community in a Semi-arid Habitat (with Emphasis on Seed-harvesting Species). Ph. D. Thesis, A.N.U., 204 pp.

Briese, D. T. & Macauley, B. J., 1977. Physical structure of an ant community in semi-arid Australia. Aust. J. Ecol. 2: 107–120.

Briese, D. T. & Macauley, B. J., 1980. Temporal structure of an ant community in semi-arid Australia. Aust. J. Ecol. 5: 121–134.

Briese, D. T. & Macauley, B. J., 1981. Food collection within an ant community in semi-arid Australia, with special reference to seed harvesters. Aust. J. Ecol. 6: 1–19.

Briese, D. T., 1982a. Resource partitioning amongst seed-harvesting ants in semi-arid Australia. Aust. J. Ecol. 7: in press.

Briese, D. T., 1982b. The effect of ants on the soil of a semi-arid saltbush habitat. Insectes Soc. 29: in press.

Brown, J. H., Davidson, D. W. & Reichman, O. J., 1979. An experimental study of competition between seed-eating desert rodents and ants. Am. Zool. 19: 1129–1143.

Brown, W. L. & Taylor, R. W., 1970. Hymenoptera: superfamily Formicoidea. In: CSIRO, The Insects of Australia, 951–959. Melbourne University Press, Melbourne.

Campbell, M. H., 1966. Theft by harvesting ants of pasture seed broadcast on unploughed land. Exp. Agric. Anim. Husb. 6: 334–338.

Davidson, D. W., 1977. Species diversity and community organization in desert seed-eating ants. Ecology 58: 711–724.

Greenslade, P. J. M., 1979. A Guide to Ants of South Australia. South Australian Museum, Adelaide, 44 pp.

Gregg, R. E., 1963. The Ants of Colorado. Univ. Colorado Press, Boulder, 792 pp.

Hall, E. A. A., Specht, R. L. & Eardley, C. M., 1964. Regeneration of the vegetation on Koonamore Vegetation Reserve 1926–62. Aust. J. Bot. 12: 205–264.

Hansen, S. R., 1978. Resource utilization and coexistence of three species of Pogonomyrmex ants in an upper Sonoran grassland community. Oecologia 35: 109–117.

Harper, J. L., 1969. The role of predation in vegetational diversity. Brookhaven Symp. Biol. 22: 48–62.

Harper, J. L. & White, J., 1970. The dynamics of plant populations. Proc. Adv. Study Inst. Dynamics Numbers Popul. (Oosterbeek, 1970): 41–63.

Hayashida, K., 1960. Studies on the ecological distribution of ants in Sapporo and its vicinity. Insectes Soc. 7: 125–162.

McGowan, A. A., 1969. Effect of seed-harvesting ants on the persistence of Wimmera rye-grass in northeastern Victoria. Exp. Agric. Anim. Husb. 9: 37–40.

Moore, C. W. E., 1953. The vegetation of the south-eastern Riverina, New South Wales, 2. The disclimax communities. Aust. J. Bot. 1: 548–567.

Morton, S. R., 1979. Diversity of desert-dwelling mammals: a comparison of Australia and North America. J. Mammal. 60: 253–264.

Morton, S. R., 1982. Granivory in the Australian arid zone: diversity of harvester ants and structure of their communities. In: Barker, W. R. & Greenslade, P. J. M. (eds.), Evolution of the Flora and Fauna of Arid Australia. Peacock, Adelaide, in press.

Noy-Meir, I., 1973. Desert ecosystems: environment and producers. Ann. Rev. Ecol. Syst. 4: 25–51.

Rogers, L. E., 1974. Foraging activity of the western harvester ant in the shortgrass plains ecosystem. Environ. Entomol. 3: 420–424.

Schumacher, A, & Whitford, W. G., 1976. Spatial and temporal variation in Chihuahuan desert ant faunas. Southwest. Nat. 21: 1–8.

Stradling, D. J., 1978. Food and feeding habits of ants. In: Brian, M. V. (ed.), Production Ecology of Ants and Termites, 81–106. C.U.P., Cambridge.

Sudd, J. H., 1967. An Introduction to the Behaviour of Ants. Arnold, London, 200 pp.

Tevis, L., 1958. Interrelations between the harvester ant *Veromessor pergandei* and some desert empherals. Ecology 39: 695–704.

Westoby, M., Cousins, J. M. & Grice, A. C., 1982. Rate of decline of soil seed populations during drought in western New South Wales, this volume.

Whitford, W. G., 1978a. Foraging by seed-harvesting ants. In: Brian, M. V. (ed.), Production Ecology of Ants and Termites, 107–110, C.U.P., Cambridge.

Whitford, W. G., 1978b. Foraging in seed-harvester ants *Pogonomyrmex* spp. Ecology 59: 185–189.

Williams, D. G., 1972. Ecological Studies on Shrub-steppe of the Western Riverina, New South Wales. Ph. D. Thesis, A.N.U., 166 pp.

Williams, O. B., 1961. Studies in the ecology of the Riverine Plain. III. Phenology of a *Danthonia caespitosa* Gaudich grassland. Aust. J. Agric. Res. 12: 247–259.

Williams, O. B., 1968. Studies in the ecology of the Riverine Plain. IV. Basal area and density changes of *Danthonia caespitosa* Gaudich in a natural pasture grazed by sheep. Aust. J. Bot. 16: 565–578.

Wilson, A. D., Leigh, J. H. & Mulham, W. E., 1969. A study of merino sheep grazing a bladder saltbush *(Atriplex vesicaria)* – cotton bush *(Kochia aphylla)* community on the Riverine Plain. Aust. J. Agric. Res. 20: 1123–1136.

CHAPTER FOUR

Restricting losses of aerially sown seed due to seed-harvesting ants

M. H. Campbell

Abstract. Seed-harvesting ants occur throughout Australia. The main species are *Monomorium, Meranoplus, Chelaner, Iridomyrmex, Chalcoponera* and *Pheidole,* the last being the most widespread. These ants become economically important only when they take seeds of aerially-sown pasture species. If conditions remain dry after sowing the ants can collect large quantities of seed and thus reduce establishment.

The most successful way to restrict losses of seeds due to ants is to coat the seed with either of the insecticides bendiocarb or permethrin before sowing. These insecticides do not reduce germination of the seed or the viability of rhizobia added to legume seed. Both insecticides have a relatively low mammalian toxicity and do not pollute the environment.

Introduction

Seed-harvesting ants take seeds into their nests and store them for future consumption; some other ants take seeds to decorate their nests. Many Australian seed-harvesting ants are widely adapted species rather than specialists; in the Northern Territory, for example, ants of the genus *Meranoplus* collect insects when seeds become scarce (Greenslade and Mott 1978). Similarly, the seed-harvesting ant *Veromessor pergandei* in the Californian desert collected 10% insects (including live termites), 68% seed and 22% inedible items when seeds became scarce after a twelve-year drought, as compared to 1% insects, 7% flower parts and 92% seed when seeds were plentiful (Tevis 1958). Even after drought the ants had little effect on the overall seed population, but since they were highly selective foragers they could produce major population changes in particular plant species. When seed was plentiful in spring, for example, 50% of the total seed taken by *V. pergandei* came from a plant species that constituted < 10% of the vegetation, the ants walking over less-preferred seed and climbing plants to take the preferred seeds.

Similarly, seed-harvesting ants can have agricultu-rally deleterious effects when introduced species are sown aerially into native pasture, because the number of seeds sown is relatively small, and the introduced seeds are generally more attractive to the ants than the existing seed population. At Katherine in the Northern Territory, Mott and McKeon (1977) found that seed-harvesting ants reduced the establishment of aerially-sown *Stylosanthes humilis,* and in Victoria, McGowan (1969) found that seed-harvesting ants hindered the regeneration and persistence of *Lolium rigidum,* even after it was well established.

Perhaps the most serious potential economic effect of seed-harvesting ants is their removal of expensive seeds of improved perennial pasture species aerially sown on seedbeds treated with herbicides (Campbell and Gilmour 1979). This renders the herbicide treatment, which can cost up to $60 per ha, ineffective because of the subsequent poor establishment of the pasture. A typical pasture mixture sown in New South Wales would be sown at a seed density of 10 kg per ha. This represents approximately 5 million seeds per ha; *V. pergandei* removes seven times this number per ha annually. Johns and Greenup (1976b) predicted that seed-harvesting ants in northern New South Wales could take from 20% to 93% of an aerial sowing of improved pasture seed depending on season of sowing.

Thus seed-harvesting ants pose a threat to the persistence of some native and introduced pasture species as well as to the establishment of aerially-sown improved species. Attempts to find a successful method for restricting the potential loss of seed due to seed-harvesting ants have included spraying the target area or ant trails with insecticides, treating sown seed with repellents or insecticides, and distributing baits containing insecticides. This chapter reviews the habits of seed-harvesting ants in Australia and methods used to restrict their depredation of sown seed.

The ants

Although the Australian ant fauna is rich, numbering approximately 2,000 species (R. Taylor, personal communication), comparatively few species harvest seeds. For example, of the 120 ant species collected in the Katherine area only seven were seed harvesters (Greenslade and Mott 1978). The few species that do harvest seed, however, appear to be abundant and widely distributed. *Pheidole* species are the most widespread seed-harvesting ants in Australia: they have been observed collecting seed in the Katherine area of the Northern Territory (Mott and McKeon 1977; Greenslade and Mott 1978); southern Queensland (Russell et al. 1967); the northwestern slopes, northern and central tablelands of New South Wales (Campbell and Swain 1973; Johns and Greenup 1976a; Campbell 1966; Campbell and Gilmour 1979); and northeastern Victoria (McGowan 1969). Identifications to species level are uncertain except in the case of *P. megacephala* Fabr. Species of *Monomorium* and *Meranoplus* also harvest seed in the Katherine area (Greenslade and Mott 1978) and in southern Queensland, where they are accompanied by seed-harvesting *Melophorus* and *Chelaner* species, and by *Iridomyrmex, Rhytidoponera* and *Polyrhachis* species which remove seed apparently to decorate their nests (Russell et al. 1967). *Iridomyrmex* and *Chalcoponera* species were also observed collecting small quantities of seed in northeastern Victoria (McGowan 1969). Most seed-harvesting ants are relatively small (Russell et al. 1967). *Pheidole* workers at Armidale, Bathurst and Orange, for example, are 3 mm long with a live weight of 0.5 mg

Campbell 1968), and can carry clover seed of 17 times their own weight.

Seed-harvesting ants in the Northern Territory prefer open disturbed sites, such as tracks and graded roadsides (Greenslade and Mott 1978). In the central tablelands of New South Wales, *Pheidole* nests are more abundant in improved than unimproved pastures, probably in response to increased seed supplies; they are also common in areas of mown turf, such as golf courses and other recreation areas, and on bare ground. Middens of soil and seed debris are characteristic of the main nest entrances.

Ant nest densities are hard to measure because the mounds are small and the entrance can be changed up to ten times a year (Tevis 1958). *Pheidole* nest densities at Orange, estimated by following ant trails from randomly distributed seed baits, range from 0 to 8 nests/m^2 with mean density 0.7 nests/m^2. Mean *Pheidole* nest density at Rockley is 0.25 nests/m^2 (Campbell 1966). These are equivalent to 7,000 and 2,500 nests/ha respectively, as compared to 15 large colonies per ha for *Veromessor pergandei* (Tevis 1958) and 27 for *Pogonomyrmex badius* (Golley and Gentry 1964). The latter colonies contained approximately 5,000 ants each, of which only 10% actively collected seeds during any two-week period. There are no corresponding data for *Pheidole*.

Seed harvesting

Australian seed-harvesting ants generally prefer small seed to large, and grass seed to legumes (Table 1). It is easier for ants to remove the palea and lemma of a grass seed than to pierce the hard seed coat of a legume. *Pheidole* species collect seed of many plant species and store them in underground granaries at various depths. Seeds may be left on the nest mound before storage, to allow weathering to soften or break the seed coat. Grass seeds are taken into the nest intact, and the palea and lemma discarded subsequently.

The activity of seed-harvesting ants is controlled primarily by temperature, and ants often show behavioural adaptations to temperature patterns (Table 2). Though unable to tolerate the temperature extremes of the Sonoran Desert (<0 °C and 56 °C respectively), *Veromessor pergandei* survives by building deep nests and foraging during periods of moderate temperature (Tevis 1958). Foraging ceases

Table 1. Seed preferences of harvester ants in relation to seed weight.

Seed species	Weight (mg per seed)	Preference (1 = highest)
Central tabelands, N.S.W. (Campbell 1966)		
Trifolium repens	0.6	4
Dactylis glomerata	0.7	1
Phalaris aquatica	1.5	3
Lolium perenne	1.9	2
Trifolium pratense	1.9	4
Trifolium subterraneum	7.5	6
Central tabelands, N.S.W. (Campbell, unpublished data)		
Phalaris aquatica	1.5	1
Medicago sativa	2.3	2
Northern tabelands, N.S.W. (Campbell and Swain 1973)		
Dactylis glomerata	0.7	2
Phalaris aquatica	1.5	1
Medicago sativa	2.3	3
Trifolium subterraneum	7.5	4
Northern tabelands, N.S.W. (Johns and Greenup 1976)		
Trifolium repens	0.6	2
Phalaris aquatica	1.5	1
Festuca arundinacea	2.3	2
Medicago sativa	2.3	4
Victoria (McGowan 1969)		
Dactylis glomerata	0.5	3
Vulpia bromoides	0.5	4
Phalaris aquatica	1.4	1
Bromus mollis	1.8	5
Lolium rigidum	2.4	1
Trifolium subterraneum	6.5	6
Hordeum leporinum	7.8	7
Queensland (Russell et al. 1967)		
Medicago sativa	2.3	1
Glycine javanica	6.1	3
Sorghum almum	6.7	2
Phaseolus atropurpureus	11.9	4
Northern Territory (Mott and McKeon 1977)		
Digitaria ciliaris	1.7	1
Stylosanthes humilis	3.2	3
Themeda australis	6.5	2

Table 2. Effect of temperature on the activity of *Veromessor pergandei* (Tevis 1958) and the percentage removal of seed by *Pheidole* species (Johns and Greenup 1976a).

Temp. (°C)	Activity of *V. pergandei*	Percentage of seed removed by *Pheidole* in 24 hours	
		Armidale	Tamworth
< 5	Ants paralyzed by cold	< 1	nm[a]
5–10	Ants moved slowly in wobbly manner	1–8	28–45
13	Ants moved steadily, 15 cm per min	15	56
20	nm	44	77
33	Ants moved quickly, 240 cm per min	nm	nm
50	Death occurred in a few seconds	nm	nm

[a] nm = not measured.

Pheidole species at two sites in northern New South Wales (Table 2). At given air temperatures rates were greater at the drier site (cf. Carroll and Janzen 1973).

Total seed removal rates in any particular area also depend on the populations of seed-harvesting ants and the sizes of their foraging territories. The *Pheidole* species studied by Mott and McKeon (1977) in the Northern Territory foraged only up to one metre from nest opening, whereas *Pheidole* species in the central tablelands of New South Wales forage several metres from nests, and *Veromessor pergandei* forages up to 40 m from the colony (Tevis 1958).

Effects on plant establishment and regeneration

In general, harvesting ants take seeds from on or above the soil surface. Their main effects are therefore on aerially- or surface-sown seed. It is not known whether ants take seed sown into prepared seedbeds in Australia, though Anslow (1958) found that ants took seed of *Setaria sphacelata* from underground sowings in Africa. Seed harvesting by ants may influence sown pastures either by reduction in the overall seed pool or by selective foraging of particular species: *Lolium rigidum* is preferentially harvested by *Pheidole* in Victoria, for example (McGowan 1969). Ant predation is more likely to influence the initial establishment of a pasture than its subsequent persistence, as shown by Mott and

at 44 °C for *V. pergandei* (Tevis 1958), and at 42 °C for *Pheidole* in the Northern Territory (Mott and McKeon 1977). On the New South Wales tablelands *Pheidole* species forage in the morning and evening in summer and at midday in winter (Campbell 1968); in the Northern Territory Mott and McKeon (1977) also found *Pheidole* foraging at night despite soil temperature as low as 20 °C. Johns and Greenup (1976a) showed that temperature accounts for 68–87% of the variance in rates of seed removal by

McKeon (1977) for introduced pastures in the Northern Territory, by Russell and Coaldrake (1965) for aerial seeding into burnt brigalow forest in Queensland, and by Johns and Greenup (1976a) for aerially sown pastures in northern New South Wales. Removal rates in the last-mentioned area ranged from < 1% per day and 25% per day in winter, at Armidale and Tamworth respectively, to 40% and 90% per day in summer. Campbell and Swain (1973) recorded a removal rate of 29% per day in winter 1969 for *Pheidole* at Armidale. Winter harvesting was also substantial at Rockley, in the central tablelands of New South Wales (Campbell 1966), despite mean temperatures of 5.5.–6.7 °C for winter months. The ants foraged at midday when temperatures approximated 11 °C. Seed removal was greatest when a long dry period followed sowing since this increases the interval between sowing and germination, and *Pheidole* prefers ungerminated to germinated seed.

Overall, substantial losses of aerially-sown seed may be expected if sowing is followed by dry weather and mean temperatures above 5 °C. Such conditions often occur in winter on the tablelands and slopes of New South Wales, and aerial sowing of pasture seed in New South Wales is generally undertaken in late autumn and winter (Campbell 1963; McDonald and Campbell 1979). As landholders in New South Wales aerially distributed an average of 2054 tonnes of seed per annum from 1969–70 to 1976–77 (Anon. 1978), and ants also take seeds sown onto recreational areas, the total cost of seed loss could be high.

Reducing seed losses

The evidence presented above shows that unless losses due to seed-harvesting ants can be reduced all aerial sowings are at risk of partial or complete failure. The simplest method of reducing losses would be to sow when seed-harvesting ants are inactive. This could be achieved by winter sowing at altitudes above 1,000 m (Johns and Greenup 1976a), but is an unreliable method as substantial losses would occur in unusually dry or warm winters (Campbell and Swain 1973; Johns and Greenup 1976a). Russell et al. (1967) found that lime pelleting reduced seed losses due to ants in Queensland, but light falls of rain, dews, frosts or mists wash the lime coating from seeds, making them attractive to ants, and even with the lime coating intact, ants will take substantial

amounts of seed in New South Wales (Campbell 1966, 1968). Repellents applied to the seed coat tend to be ineffective because they break down within a week on the soil surface (C. R. Wallace, personal communication). Russell et al. (1967) proposed the development of more effective repellents from compounds resembling myrmecoid secretions, but these have yet to be isolated. Seed losses can also be reduced by distributing seeds with baits containing the formicide mirex together with substances attractive to ants (Campbell 1977), but many non-target ants are also killed. This treatment has been discontinued because kepone, an analogue of mirex, was found damaging to human health and the environment.

Insecticides applied to the seed coat have been used for some time to reduce losses due to seed-harvesting ants. Those used with success include dieldrin (Anslow 1958), lindane (Champ and Sillar 1961), and chlordane (Anon. 1964). Dieldrin coating increased mean establishment rates of three legumes and grasses by factors of 2 to 5, irrespective of sowing season, in pastures on the central tablelands of New South Wales (Campbell 1966). *Pheidole* workers took treated seed into their nests for approximately one hour after sowing, and foraging activity fell markedly soon after, remaining low for three to four weeks. Between 10 and 80 seeds were sufficient to inactivate a nest. Many dead ants were thrown out of the affected nests; some nests recovered, but others were completely inactivated. Insecticide tests in Queensland did not show effective reduction in seed losses, however (Russell et al. 1967), owing presumably to differences in the ant fauna or to the low number of seeds used per plot (10) in their experiment.

Dieldrin treatment of seed was discontinued in the early 1970's because of its toxic effect on animals and its persistence in animal tissue. In 1978 a search was begun for new insecticides that could be used to reduce seed losses due to ants without harming the environment. Bendiocarb and permethrin were found to be effective (Campbell and Gilmour 1979): treatment reduced removal rates from over 150 seeds/nest/day to only 5 seeds/nest/day over a 14-day period. Bendiocarb has a similar effect on ants to dieldrin but permethrin acts more as a repellent (Campbell and Gilmour 1979). These experiments employed seed depots placed close to nest entrances, so two further experiments were carried out in autumn 1980, using treated and untreated seeds broadcast on the soil surface in simulated aerial seedings. In the first experiment, selected one m² areas supporting one or more ant nests were sown; in the second, areas of 49 m² were sown but only the central one m² observed. The numbers of seeds taken by *Pheidole* workers were recorded in each case, using seven replications of four treatments in the first experiment and four replications of five treatments in the second. The results show that permethrin and bendiocarb are effective ($p < 0.05$) in restricting losses due to ants even when the seeds are scattered on the soil surface (Tables 3 and 4). The apparent ineffectiveness of insecticides in reducing losses of *M. sativa* in the second experiment (Table 4) is due to reduced density of ant nests on the control relative to the other treatments; losses per nest were over twice as high for the control as for any other treatment. In both experiments the removal of treated seeds was substantial owing to the high nest density on the quadrats (2.5 to 5.0 per m² compared to 0.7 nests per

m² in the surrounding area), seed theft by ants from outside the quadrats, and the intense foraging activity under the prevailing warm, dry conditions (mean air temperature 16 °C and 12 °C in experiments 1 and 2, respectively). Losses of treated seed in the field will be lower but could still be considerable, so more effective treatments are still required.

Permethrin and bendiocarb are less toxic than dieldrin (Table 5), and are effective at lower rates (Campbell and Gilmour 1979), so the chances of environmental damage are reduced. Permethrin and bendiocarb wettable powders do not reduce germination of *M. sativa* or *P. aquatica* or survival of *Rhizobium meliloti* or *R. trifolii*, so they can be applied to legume seed coats in the normal lime-pelleting process. The miscible-oil formulation of permethrin reduces both the rate and final germination percentage of *P. aquatica* seed, but this is due to the oil rather than the permethrin (Campbell and Gilmour 1979). Miscible-oil formulations of perme-

Table 3. Effect of coating seed with insecticide on the removal of simulated aerially-sown seed by ants (experiment 1).

Seed treatment		Ant nests (per m²)	Seed remaining after 21 dry days			
Insecticide	Rate[a]		*M. sativa*		*P. aquatica*	
			% of seed sown[b]	No. per m²	% of seed sown	No. per m²
Permethrin	0.16	3.9	26 a	47	24 a	66
Bendiocarb	0.16	4.0	15 b	27	6 b	17
Chlorpyrifos	0.16	2.7	6 c	10	< 1 c	2
Control	–	3.9	7 c	13	< 1 c	< 1

[a] In kg a.i. per 100 kg seed.
[b] Means in columns not followed by the same letter differ significantly ($p < 0.05$). Analyses are based on the angular transformation.

Table 4. Effect of type and rate of insecticide on the removal of simulated aerially-sown seed by ants (experiment 2).

Seed treatment		Ant nests (per m²)	Seed remaining after 11 dry days			
Insecticide	Rate[a]		*M. sativa*		*P. aquatica*	
			% of seed sown[b]	No. per m²	% of seed sown	No. per m²
Permethrin	0.32	4.5	53 a[a]	98	53 a	145
	0.16	3.5	56 a	104	46 ab	126
Bendiocarb	0.32	3.7	48 a	89	38 b	104
	0.16	5.0	39 a	72	31 b	85
Control	–	5	32 a	59	6 c	16

[a] In kg a.i. per 100 kg seed.
[b] Means in columns not followed by the same letter differ significantly ($p < 0.005$). Analyses are based on the angular transformation.

Table 5. Toxicity of insecticides.

	Oral LD 50	Dermal LD 50
	(mg/kg body weight, male rat)	
Permethrin	1,479–20,000	4,000
Bendiocarb	68	> 1,000
Dieldrin	46	98

thrin and bendiocarb are easier and safer to use in seed coating than are wettable powders, but the solvent oil used must be chosen so that seed germination is not affected.

Conclusions

Although seed-harvesting ants are widespread in Australia and can take large amounts of aerially-sown seed, little is known of their ecology and they can rarely be identified to species level. At the moment the only proven method of reducing the amount of seed taken by these ants is to treat the seeds with insecticides. This is not completely effective as some seeds are taken despite the treatment, and the insecticides may have additional unknown deleterious effects.

Thus there is an urgent practical need for a complete Australia-wide study of the ecology and taxonomy of seed-harvesting ants, to assist in the development of more effective means to reduce economic losses due to seed-harvesting ants.

References

Anon, 1964. Soil and pasture research on the northern tablelands of New South Wales. Agricultural Research Liaison, C.S.I.R.O., Melbourne.

Anon, 1978. Crop and pasture statistics 1976–77 season. Australian Bureau of Statistics, Cat. 7301.1.

Anslow, R. C., 1958. A note on an improved technique in establishing *Setaria sphacelata* from seed. Emp. J. Exp. Agric. 26: 55–57.

Campbell, M. H., 1963. Establishment of pasture species on unploughed ground. Third Aust. Grassl. Conf., Paper 4-3.

Campbell, M. H., 1966. Theft by harvesting ants of pasture seed broadcast on unploughed land. Aust. J. Exp. Agric. Anim. Husb. 6: 334–338.

Campbell, M. H., 1968. Cops and robbers. Chiasma 6: 93.

Campbell, M. H., 1977. No remedy for seed-stealing ants. Agric. Gaz N.S.W. 88: 44.

Campbell, M. H. & Gilmour, A. R., 1979. Reducing losses of surface-sown seed due to harvesting ants. Aust. J. Exp. Agric, Anim. Husb. 19: 706–711.

Campbell, M. H. & Swain, F. G., 1973. Factors causing losses during the establishment of surface-sown pastures. J. Range Manage 26: 355–359.

Carroll, C. R. & Janzen, D. H., 1973. Ecology of foraging by ants. Ann. Rev. Ecol. Syst. 4: 231–259.

Champ, B. R. & Sillar, D. I., 1961. Pellet your buffel seed and thwart ants. Queensl. Agric. J. 87: 583.

Golley, F. B. & Gentry, J. B., 1964. Bioenergetics of the southern harvester ant Pogonomyrmex badius. Ecology 45: 217–225.

Greenslade, P. J. M. & Mott, J. J., 1978. Ants (Hymenoptera, Formicidae), of native and sown pastures in the Katherine area, N.T., Australia. Second Australasian Conf. Grassl. Invert. Ecol., Session 5. Palmerston North, N.Z.

Johns, G. G. & Greenup, L. R., 1976a. Pasture seed theft by ants in northern New South Wales. Aust. J. Exp. Agric. Anim, Husb. 16: 249–256.

Johns, G. G. & Greenup, L. R., 1976b. Predictions of likely theft by ants of oversown seed for the northern tablelands of New South Wales. Aust. J. Exp. Agric. Anim. Husb. 16: 257–264.

McDonald, W. J. & Campbell, M. H., 1979. Replacing thistles with perennial pastures on non-arable land. Wool Technology and Sheep Breeding 27 (3): 31–34.

McGowan, A. A., 1969. Effect of seed-harvesting ants on the persistence of Wimmera ryegrass in pastures in northeastern Victoria. Aust. J. Exp. Agric. Anim. Husb. 9: 37–40.

Mott, J. J. & McKeon, G. M., 1977. A note on the selection of seed types by harvester ants in northern Australia. Aust. J. Ecol. 2: 231–235.

Russell, M. J. & Coaldrake, J. E., 1965. C.S.I.R.O. Division of Tropical Pastures, Annual Report 1964.

Russell, M. J., Coaldrake, J. E. & Sanders, A. M., 1967. Comparative effectiveness of some insecticides, repellents and seed-pelleting devices in the prevention of ant removal of pasture seeds. Trop. Grassl. 1: 153–166.

Tevis, L., 1958. Interrelations between the harvester ant *Veromessor pergandei* (Mayr.) and some desert ephemerals. Ecology 39: 695–704.

CHAPTER FIVE

Seed removal by ants in the mallee of northwestern Victoria

Alan Andersen

Abstract This study examines seed removal by ants in the mallee region of northwestern Victoria as one mechanism by which the distributions of ant and plant communities may influence each other. Marked differences in vegetation of two adjacent sites (mallee and heath) are matched by pronounced differences in their associated ant faunas. Harvester ants rapidly removed a wide variety of seeds placed at baiting stations. The ants showed considerable seed-species selectivity and marked spatial variability, indicating that they may play an important role in plant distribution. The distributions of particular harvester ants appear to be determined by the availability of preferred seeds. The activities of elaiosome-collecting ants are discussed briefly.

Introduction

It has frequently been reported that ants may collect large quantities of seeds from the soil surface, usually to obtain the food reserves they contain (Ashton 1979; Briese and Macauley 1981; Brown et al. 1979; Majer 1980; Tevis 1958; Willard and Crowell 1965), or for the special ant-attracting appendages (elaiosomes) some seeds possess (Berg 1972, 1975; Culver & Beattie 1978, 1980; Harper et al. 1970; Horvitz & Beattie 1980; Majer et al. 1979; O'Dowd & Hay 1980).

Initial work on seed removal by ants in Australia was inspired primarily by economic interests, owing to their interference with agricultural seeding operations (Campbell 1966; McGowan 1969; Mott & McKeon 1977; Russell et al. 1967; Smith & Atherton 1944). It is now recognized that seed-removing ants may play an important role in Australian deserts (Morton 1982), where ephemeral vegetation often produces vast quantities of seeds. Recent work suggests they may also be important elements in the more complex ecosystems of semi-arid and temperate Australia (Ashton 1979; Briese & Macauley 1981; Hodgkinson et al. 1980; Majer 1980; Shea et al. 1979).

The removal of seeds by ants provides a mechanism by which the distributions of plant and ant communities may influence each other: ants may influence the distribution of plants by dispersing or eating their seeds, and plants may influence the distribution of ants by providing them with a food source. This study examines such influences in two adjacent plant communities in the mallee region of northwestern Victoria. Its objectives are to describe and compare the vegetation, the ant faunas, and seed removal by ants in the two communities.

The study site (Fig. 1) is located just west of the northwestern shore of Lake Albacutya in northwestern Victoria. It lies inside the southern extension of Wyperfeld National Park and on the eastern fringe of the Big Desert. The area consists of jumbled parabolic dunes of white Pleistocene siliceous sands overlying Tertiary marine deposits (Hills 1975; Lawrence 1966). It has a semi-arid climate (Cheal et al. 1979), with hot summers (mean daily maximum and minimum temperatures 30 °C and 13 °C, respectively), cool to mild winters (15 °C and 4 °C), and an average annual rainfall of 400 mm with a slight winter peak. The dominant vegetation is a typical mallee formation of low, lignotuberous multi-stemmed eucalypts forming a more or less continuous canopy at a height of

Buckley, R. C. (ed.), Ant-plant interactions in Australia.
© 1982, Dr W. Junk Publishers, The Hague. ISBN 90 6193 684 5.

31

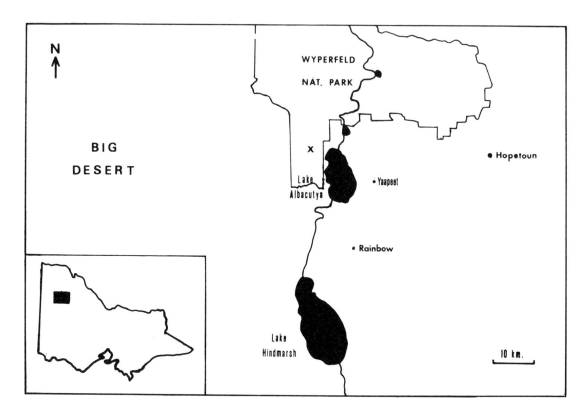

Fig. 1. Location of study site (X) in northwestern Victoria.

5–6 m. Three mallee species are found in the area, in both mixed and pure stands: *E. incrassata, E. dumosa* and *E. foecunda* (nomenclature follows Willis 1970, 1972 throughout). The understorey consists mainly of sclerophyllous shrubs, and includes *Baeckea behrii, B. crassifolia, Leptospermum laevigatum* var *minus, Calytrix tetragona,* and *Melaleuca uncinata* (Myrtaceae); *Lasiopetalum behrii* and *L. baueri,* Sterculiaceae); *Bertya mitchellii* (Euphorbiaceae); *Phebalium bullatum* (Rutaceae); *Hibbertia stricta* (Dilleniaceae); *Aotus ericoides* (Papilionaceae); *Dodonaea bursariifolia* (Sapindaceae) and *Prostanthera aspalathoides* (Labiatae). Porcupine grass *(Triodia irritans,* Gramineae) is often common on coarser sands. The mallee frequently gives rise to an open heath formation toward the top of sandhills. The heath includes several shrub species from the mallee understorey, together with *Banksia ornata* and *Hakea vittata* (Proteaceae); *Acacia calamifolia* (Mimosaceae); *Phyllota pleurandroides* (Papilionaceae), *Casuarina pusilla* (Casuarinaceae); and *Brachyloma ericoides, B. daphnoides, Astroloma conostephioides* and *Leucopogon cordifolius* (Epacri-

daceae). *Callitris verrucosa* (Cupressaceae) may also be present.

I chose an area for my study sites where a fire access track follows the slope of a sandhill and marks a discontinuity between the mallee formation below and the heath formation above. The dominant mallee eucalypt is *E. incrassata,* with *E. dumosa* also present. *Triodia irritans* is completely absent and *Callitris verrucosa* is only a minor element. An occasional stunted eucalypt is present in the heath. The study was conducted between January and June 1980.

Vegetation

The vegetation of each site was described in six 5 m $\times$ 5 m plots randomly located inside 50 m $\times$ 25 m grids (Fig. 2). Within each plot all plant species were recorded, together with the number of individuals, mean height and percent cover of each species. The results show marked differences between sites in the numbers of individuals of each species (Fig. 3).

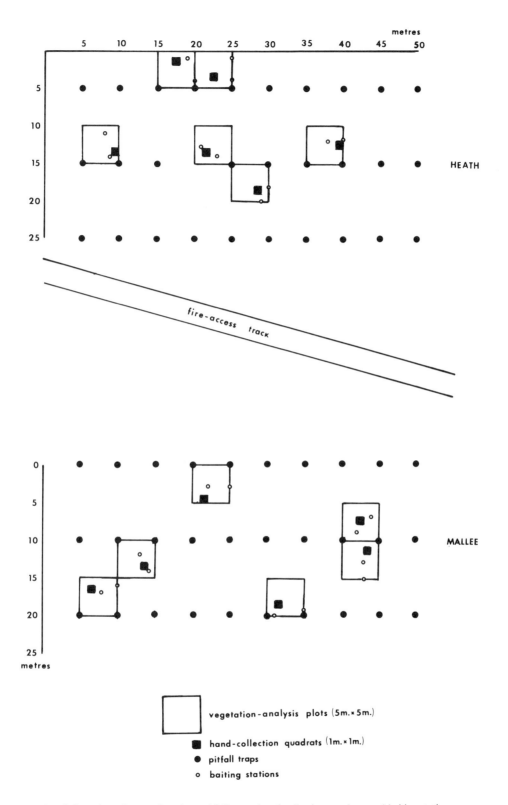

Fig. 2. Location of vegetation plots, pitfall traps, hand collection quadrats and baiting stations.

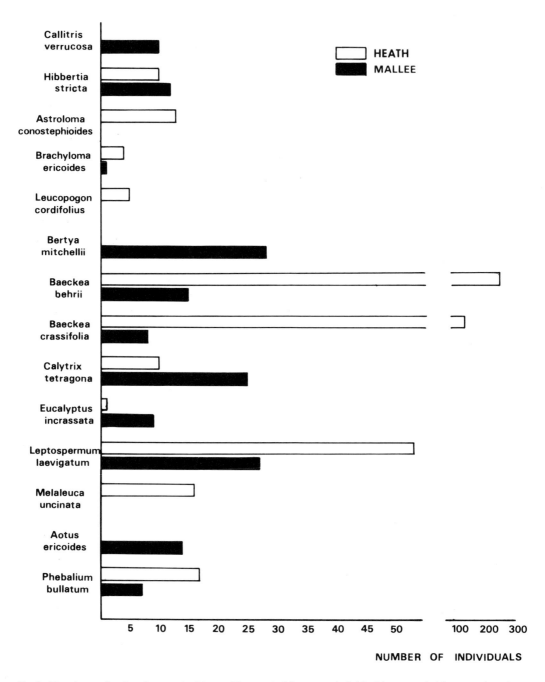

Fig. 3. Abundance of major plant species (those with a total of 5 or more individuals) as recorded in vegetation plots.

Baeckea behrii and *B. crassifolia* are by far the most common plants in the heath, whereas *Bertya mitchellii, Leptospermum laevigatum* var *minus* and *Calytrix tetragona* predominate in the mallee. Simultaneous analyses of abundance, height and cover data, using the T-square statistic (Anderson 1960),

demonstrate statistically significant differences in plant species abundance between the two sites (Table 1). Of the 14 most abundant species, only *Hibbertia stricta* and *Calytrix tetragona* show no marked differences between sites.

Analyses of auger core samples to 1.5 m depth

Table 1. Differences in the vegetation of the two sites. Only plant species with a total of 5 or more individuals recorded are considered. For each species it is indicated whether its distribution favours the mallee (M) or the heath (H). Results of chi-square analyses.

Exclusive to a site	Significant difference between sites ($p < 0.05$)	No significant difference between sites ($p > 0.05$)
Callitris verrucosa (M)	Eucalyptus incrassata (M)	Hibbertia stricta
Bertya mitchelii (M)	($F_{(3,6)} = 10.0$)	($F_{(3,6)} = 0.5$)
Aotus ericoides (M)	Phebalium bullatum (H)	Calytrix tetragona
Astroloma conostephioides (H)	($F_{(3,6)} = 5.3$)	($F_{(3,6)} = 2.8$)
Leucopogon cordifolius (H)	Brachyloma ericoides (H)	
Melaleuca uncinata (H)	($F_{(3,6)} = 33.7$)	
	Baeckea behrii (H)	
	($F_{(3,6)} = 9.6$)	
	Baeckea crassifolia (H)	
	($F_{(3,6)} = 10.6$)	
	Leptospermum laevigatum (H)	
	($F_{(3,6)} = 6.6$) var minus	

show no differences in texture or particle size between the soils of the two sites. Soil suitability for nest construction therefore does not contribute to any differences between the ant faunas of the two sites.

Ant faunas

The ants of the two sites were described from the results of pitfall traps and hand collections. Thirty pitfall traps (plastic drinking cups containing an ethanol-glycerol mixture, buried with their rims flush to the soil surface) were arranged systematically at each site (Fig. 2). The traps were operated for periods of 48 hours during summer, autumn and winter, after allowing for 'digging-in' periods (Greenslade 1973) of several days. Hand collections were conducted within permanent 1 m × 1 m quadrats located randomly in each vegetation plot (Fig. 2). Ants were collected from each quadrat for five minutes during each 2-hour period of the day (00–0200 hrs, 02–0400 hrs, . . . etc.), in summer, autumn and winter. The ants were identified to genus, and the different species of each genus distinguished and numbered. Species designations were checked by Dr R. W. Taylor.

A total of 80 species representing 24 genera were recorded from the sites, 59 of these from the heath and 64 from the mallee (Table 2). The most important genera were *Iridomyrmex* (12 species), *Melophorus* (15 species), *Camponotus* (11 species) and *Stigmacros* (9 species). Species richness and popula-

Table 2. Ant genera and number of species per genus recorded from study sites.

Sub-family Genus	Number of species		
	Heath	Mallee	Total
MYRMICINAE			
Adlerzia Forel	1	1	1
Aphaenogaster Mayr	1		1
Chelaner Emery	2	3	3
Colobostruma Wheeler	1	1	1
Crematogaster Lund	2	2	2
Epopostruma Forel	1	1	2
Meranoplus F. Smith	3	3	3
Mesostruma Lund		1	1
Monomorium Mayr	4	5	5
Pheidole Westwood	2	2	2
Tetramorium Mayr	1	1	1
PONERINAE			
Cerapachys F. Smith	1		1
DOLICHODERINAE			
Dolichoderus Lund		2	2
Froggatella Forel		1	1
Iridomyrmex Mayr	10	10	12
FORMICINAE			
Acropyga Roger		1	1
Camponotus Mayr	8	8	11
Melophorus Lubbock	12	12	15
Notoncus Emery		2	2
Paratrechina Motschoulsky	1	1	1
Plagiolepis Mayr	1	1	1
Polyrhachis F. Smith	1	1	1
Prolasius Forel		1	1
Stigmacros Forel	7	5	9
Total	59	64	80

Table 3. Distribution of the 34 most common ant species (those with a total of more than 15 individuals recorded) across sites. A species is said to be exclusive to a site if more than 95% of its individuals were recorded there. A species showed a strong preference for a particular site if between 75% and 95% of its individuals were found there. Combined data from hand collections and pitfall traps.

Exclusive (> 95%)		Strong preference (75–95%)		No marked
Heath	Mallee	Heath	Mallee	preference (< 75%)
Monomorium sp.1	*Monomorium* sp.3	*Monomorium* sp.4	*Iridomyrmex* sp.1	*Crematogaster* sp.1
Iridomyrmex sp.2	*Chelaner* sp.3	*Pheidole* sp.1	*Iridomyrmex* sp.9	*Crematogaster* sp.2
Melophorus sp.12	*Camponotus* sp.3	*Iridomyrmex* sp.5	*Melophorus* sp.15	*Meranoplus* sp.1
Paratrechina sp.	*Melophorus* sp.6	*Melophorus* sp.1	*Melophorus* sp. 9	*Monomorium* sp.2
Stigmacros sp.8	*Notoncus* sp.2	*Melophorus* sp.2		*Iridomyrmex* sp.3
	Prolasius sp.	*Stigmacros* sp.1		*Iridomyrmex* sp. 4
	Stigmacros sp.7			*Iridomyrmex* sp.7
				Camponotus sp.1
15% in all	21% in all	18% in all	12% in all	*Camponotus* sp.5
				Melophorus sp.5
35% in all		29% in all		*Melophorus* sp.14
				Polyrhachis sp.
				35% in all

tion densities showed extreme seasonality, and individual species displayed distinctive diel patterns of foraging activity (Andersen 1980).

Remarkable differences are apparent in the composition of the ant faunas of the two sites. Of the 80 species recorded overall, only 43 were found at both sites. The distribution of the 34 most abundant species across sites is given in Table 3. Thirty-five percent of these species were found exclusively at a particular site and a further 29% showed a strong site preference. These differences extended to the dominant ants of the two communities (Andersen 1980). There were only two instances where a particular species attained dominant or sub-dominant status at both sites concurrently *(Iridomyrmex* sp.7 in autumn and *Iridomyrmex* sp.4 in winter). The most abundant ant in the mallee, *Monomorium* sp.3, was completely absent from the heath. Spearman rank correlations and chi-square analyses (Table 4) show highly significant differences in the composition of the two communities during all seasons.

Seed removal by ants

Extent of seed removal

I investigated the extent of seed removal by ants from the soil surface by placing seeds at baiting stations (Fig. 2), and monitoring their disappearance over three days. Stations were surrounded by birdwire

Table 4. Comparison of ants from mallee and heath. Results of Spearman rank correlation and chi-square analyses.

	Correlation	Chi-square
Summer	$r_{s(N=23)} = -0.047$	$\chi^2_{(22)} = 3611$
	$p > 0.05$	$p < 0.001$
Autumn	$r_{s(N=12)} = 0.154$	$\chi^2_{(11)} = 452$
	$p > 0.05$	$p < 0.001$
Winter	$r_{s(N=25)} = 0.291$	$\chi^2_{(9)} = 323$
	$p > 0.05$	$p < 0.001$

cages to prevent removal by granivorous birds or rodents. The seeds of 12 species were used: *Callitris verrucosa, Casuarina pusilla, Eucalyptus dumosa, E. foecunda, E. incrassata, Hakea vittata, Leptospermum laevigatum* var *minus, Melaleuca uncinata, Aotus ericoides, Hibbertia stricta, Lasiopetalum behrii* and *Phebalium bullatum,* all of which are native to the study area. The seeds of the last four species possess elaiosome-like structures. This experiment was conducted in summer and repeated in winter, except that in the latter case the seeds of *Hibbertia stricta* and *Melaleuca uncinata* were replaced respectively by seeds of *Acacia saligna* (which bear prominent elaisome-like structures) and *Eucalyptus radiata,* both of which are foreign to the area.

The results of these experiments (Table 5) indicate that seed removal by ants involves a wide variety of

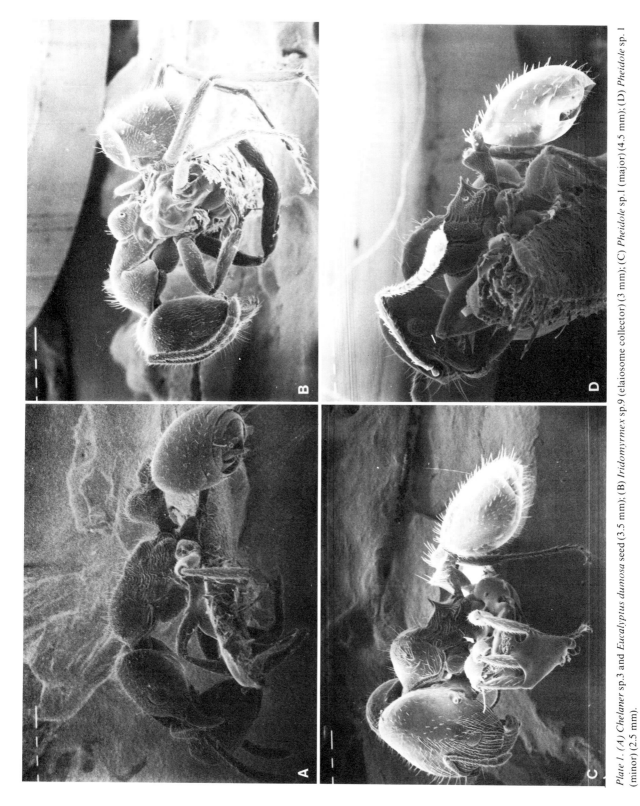

Plate 1. (A) Chelaner sp.3 and Eucalyptus dumosa seed (3.5 mm); (B) Iridomyrmex sp.9 (elaiosome collector) (3 mm); (C) Pheidole sp.1 (major) (4.5 mm); (D) Pheidole sp.1 (minor) (2.5 mm).

37

Table 5. Cumulative removal of seeds from baiting stations. Figures represent the number of seeds removed after 1, 2 and 3 days. A total of 15 seeds of each species was placed at 3 baiting stations within each study site during summer and winter. Ae – *Aotus ericoides*; Cv – *Callitris verrucosa*; Cp – *Casuarina pusilla*; Ed – *Eucalyptus dumosa*; Ef – *E. foecunda*; Ei – *E. incrassata*; Hv – *Hakea vittata*; Hs – *Hibbertia stricta*; Lb – *Lasiopetalum behrii*; Ll – *Leptospermum laevigatum var minus*; Mu – *Melaleuca uncinata*; Pb – *Phebalium bullatum*; As – *Acacia saligna*; and Er – *Eucalyptus radiata*.

Heath

	Ae	Cv	Cp	Ed	Ef	Ei	Hv	Hs	Lb	Ll	Mu	Pb	Total
Summer													
Day 1	0	0	3	11	15	10	5	5	2	2	1	4	58(32%)
Day 2	0	5	6	14	15	10	5	7	2	4	2	5	75(42%)
Day 3	0	5	7	14	15	10	6	7	8	4	5	7	88(49%)
								As			Er		
Winter													
Day 1	15	0	5	5	13	10	0	15	5	1	5	5	79(44%)
Day 2	15	0	10	15	15	10	1	15	5	5	5	6	102(57%)
Day 3	15	2	11	15	15	10	3	15	6	5	5	8	110(61%)

Mallee

	Ae	Cv	Cp	Ed	Ef	Ei	Hv	Hs	Lb	Ll	Mu	Pb	Total
Summer													
Day 1	0	0	0	1	8	5	2	3	8	8	5	12	52(29%)
Day 2	1	0	1	5	10	6	4	6	10	11	6	15	75(42%)
Day 3	3	0	1	8	10	6	4	6	10	11	7	15	81(45%)
								As			Er		
Winter													
Day 1	13	0	5	10	13	10	0	15	4	0	1	1	72(40%)
Day 2	13	0	7	10	15	13	0	15	5	0	2	1	81(45%)
Day 3	13	0	7	10	15	13	0	15	6	0	3	1	83(46%)

plants, including both native and introduced species, and seeds with and without elaiosome-like structures. I have witnessed ants removing the seeds of each species tested, so that it is possible that most, if not all, species of plants in the area are affected. Up to 61% of all seeds placed at baiting stations were removed within three days, most during the first day. Considerable species selectivity was displayed by ants toward seeds. Some seeds (e.g. those of *Callitris verrucosa*) were rarely removed, whereas others (e.g. *Eucalyptus* spp.) were removed very rapidly. Seeds bearing elaiosome-like structures were not removed at a greater rate than others ($t(140) = 0.656, p > 0.05$).

The results of a two-way analysis of variance on day 3 data indicate that there were no significant differences between overall summer and winter rates of seed removal ($F(1,44) = 0.413, p > 0.05$), or between total rates of removal in the mallee and heath ($F(1,44) = 0.850, p > 0.05$). The winter removal rate in the heath, however, was significantly higher than that in the mallee: 61% vs. 46% ($T = 4.5$, $p < 0.05$, Wilcoxon matched pair test). This was probably due to the greater abundance of a harvesting species of *Pheidole* in the heath at this time (Andersen 1980).

Ant species involved

To determine the ant species responsible for seed removal I placed seeds at the baiting stations and patrolled them systematically for several days. This was done in summer, autumn and winter. The results of these experiments are summarized in Table 6. A

total of eight seed-removing species was recorded: seven of these *(Pheidole* sp.1, *Chelaner* sp.3, *Meranoplus* spp. 1 and 2, *Chelaner* spp. 1 and 2, and *Monomorium* sp.2) are true seed harvesters, since they remove non-appendaged seeds, presumably for their food reserves, and the remaining species *(Iridomyrmex* sp.9) is an elaiosome collector, since it removes only appendaged seeds.

The three most important seed-removing species *(Pheidole* sp.1, *Chelaner* sp.3 and *Iridomyrmex* sp.9: Plate 1) showed marked seasonal and site preferences (Andersen 1980). *Pheidole* sp.1 was the only species active during all seasons. It is a prolific generalist harvester (removing almost all seeds offered) common to both sites, but particularly abundant in the heath. *Chelaner* sp.3 was a major seed harvester in the mallee site during winter. Preference experiments indicate that it removes eucalypt seeds exclusively. *Iridomyrmex* sp.9 was an abundant elaiosome collector during the colder months (its activity was restricted to this period), particularly in the mallee. All three species formed foraging columns from their nests to the baiting stations.

Species of *Pheidole* are recognized as major seed harvesters throughout Australia (Briese & Macauley 1981; Campbell 1966; Johns & Greenup 1976; Mott & McKeon 1977; Russell et al. 1967; Smith & Atherton 1944) and other parts of the world (Bequaert 1922; Creighton 1966; Creighton & Creighton 1959; Kemp 1951). It has been suggested that the pronounced dimorphism shown by *Pheidole* workers (Plate 1) may be related to the seed-harvesting habit, with the major workers, possessing large heads and powerful jaws, acting as 'nutcrackers' (Creighton 1966). However it appears that in most cases their main function is nest defence (Briese 1974; Creighton & Creighton 1959). Species of *Meranoplus*, *Monomorium* and *Chelaner* are also recognized as seed harvesters (Ashton 1979; Briese & Macauley 1981; Greenslade 1979; Russell et al. 1967; Smith & Atherton 1944).

Spatial variability of seed removal

When conducting the baiting experiments described above, it became apparent that there was considerable spatial variability in the activities of seed-removing ants: seeds were consistently removed at high rates from some stations, only sporadically removed from others, and in some cases never removed. To

Table 6. Summary of seed-removing species of ants.

Species	Notes
Pheidole sp.1	Prolific seed harvester throughout study period at both sites, particularly in the heath. Generalist.
Chelaner sp.3	Major harvester of eucalypt seeds in the mallee during winter. Specialist.
Iridomyrmex sp.9	Abundant elaiosome collector during winter, particularly in the mallee.
Meranoplus sp.1 *Meranoplus* sp.2 *Clelaner* sp.1 *Chelaner* sp.2 *Monomorium* sp.2	Minor seed harvesters during different times of the study period.

investigate this matter further, I established six series of baiting stations located in and adjacent to both sites. Each series consisted of 10 stations located within 10 m of each other. I baited the stations with eucalypt seeds and recorded the ant species removing them after 12, 24 and 48 hours.

The results of this experiment (Table 7) show that seed removal by ants is remarkably variable within a comparatively small area of apparently uniform vegetation. After 12 hours, the number of stations visited by harvester ants in each series of 10 stations ranged from zero to ten in the mallee, and from one to eight in the heath. The situation had not altered appreciably after 48 hours. This spatial variability suggests that there are subtle micro-habitat differences within each vegetation community. It indicates that seeds may remain on the soil surface for long periods in some areas, while being removed rapidly from others.

Fate of seeds removed by ants

Many harvester ants store seeds in special nest granaries (Brown et al. 1979). Seed storage is particularly prevalent in desert ants (Carroll & Janzen 1973), where ephemeral vegetation often produces seeds in large and unpredictable pulses. Cache hoarding is viewed as an adaptation to wait out periods between successive seed productions (Janzen 1971) and thus buffers the ants against an unpredictable environment (Briese & Macauley 1981). It may also allow ants to remain within their nests during cold periods Cole (1934a).

Species of *Pheidole* are known to store seeds in granaries within their nests (Creighton 1966; Smith & Atherton 1944), mostly in the upper 15 cm

(Briese & Macauley 1981). I investigated seed storage in the nests of *Pheidole* sp.1 and *Chelaner* sp.3 by taking core samples over nest entrances using a 700 ml tin can (14.5 cm × 8.5 cm). After sampling 14 nests I concluded that these species do not maintain extensive granaries, unless they are always located more than 15 cm below the surface. Seed storage by ants may not be necessary in the mallee, since seeds (particularly those of eucalypts) are available throughout the year (Andersen 1980). Granaries do not appear to be maintained by the main seed-harvesting ants in *Eucalyptus* forests elsewhere (Ashton 1979; Majer et al. 1979), although I have observed seed storage by *meranoplus* sp. in eucalypt forests of the Cathedral Ranges in central Victoria.

Seeds storage has also been reported in elaiosome-collecting ants (Culver & Beattie 1980; Majer et al. 1979), but it appears that seeds are usually returned to the surface once the elaiosomes have been removed (Berg 1975; Bullock 1974; O'Dowd & Hay 1980). To investigate the fate of seeds collected by *Iridomyrmex* sp. 9, I placed 100 *Acacia saligna* seeds near the nest entrance of a large colony. All seeds were collected and taken into the nest by workers within 20 minutes. Seven days later I took a core sample of the nest in the manner described above. Forty-three seeds were recovered from this sample. The elaiosome-like structures of three of these had been completely removed and pieces of most others appeared to be missing. There was no damage evident to the seeds themselves, nor was there any evidence that seeds were returned to the surface (although this may have occurred at a later date). Since nearly half the seeds collected by workers were recovered undamaged within 15 cm of the surface, it is possible that some were located in positions favou-

Table 7. Spatial variability of seed removal. Number of stations (out of 10) at which ants were observed removing seeds. Series 1 – located within study sites. Series 2 and 3 – located immediately on either side of study sites.

Site	Heath												Mallee											
Series	1			2			3			Total			1			2			3			Total		
Baiting period (hrs)	12	24	48	12	24	48	12	24	48	12	24	48	12	24	48	12	24	48	12	24	48	12	24	48
Pheidole sp.1	7	7	7	7	7	7	1	3	3	15	17	17	–	–	–	7	7	7	–	–	–	7	7	7
Chelaner sp.3	2	2	2	–	–	–	–	–	–	2	2	2	–	–	2	8	8	8	–	–	1	8	8	11
Chelaner sp.2	–	–	–	–	–	–	–	–	1	–	–	1	–	–	–	–	–	–	–	–	1	–	–	1
Total No. of stations at which seeds were removed	8	8	8	7	7	7	1	3	4	16	18	19	0	0	2	10	10	10	0	0	2	10	10	14

rable for germination. The rapid removal of *Acacia* seeds by ants may further increase their chances of success by reducing predation by other granivores (cf. O'Dowd & Hay 1980). It may also afford the seeds protection from fire and help build up soil seed reserves during inter-fire periods (Berg 1975).

Discussion

The marked differences between the vegetation of the adjacent mallee and heath sites are matched by striking differences in the associated ant faunas. This is consistent with the common finding that distinctive ant assemblages are associated with different vegetation types (Ashton 1979; Brian 1964; Brown 1959; Cole 1932a, 1934b; Greenslade 1971; Greenslade & Greenslade 1977; Hayashida 1960; Majer 1972; Morton 1982; Sanders 1970; Talbot 1934; Weber 1943; Yasuno 1963).

An investigation of seed removal by ants provides one explanation of how the distributions of plants and ants may influence each other. Many workers have shown that vegetation determines the distribution and success of harvester ants (Morton 1982; Sharpe & Barr 1960) by influencing the availability of preferred seeds (Briese 1974; Brown et al. 1979; Bernstein 1974; Cole 1932b; Davidson 1977; Majer et al. 1979; Smith & Atherton 1944; Whitford 1976; Whitford & Ettershank 1975). Seed availability appears to have influenced the distribution of mallee harvester ants. *Chelaner* sp. 3, apparently a specialist harvester of mallee eucalypt seeds, was not found in the heath where eucalypts are absent. By contrast, *Pheidole* sp. 1 is a generalist seed harvester and was found in both habitats. A brief survey of the distributions of these ants in other vegetation communities of the region *(Eucalyptus camaldulensis* and *E. largiflorens* woodlands, *Leptospermum laevigatum* var *minus* shrubland, open grassland and further mallee stands) confirms that *Chelaner* sp. 3 is restricted to areas of mallee eucalypts, whereas *Pheidole* sp. 1 is found in nearly all habitats. *Chelaner* sp. 3 is also absent from recently burned areas where there has been no input of eucalypt seeds for several years. Seed availability may have also influenced seed preferences. All harvester ants recorded in the study showed a distinct preference for eucalypt seeds over all others – and these are precisely the seeds that are most likely to be both plentiful and available throughout the year.

The notion that seed availability modifies the composition and preferences of harvester ant faunas is supported by studies of eucalypt seed removal in other mallee stands of the study area (Andersen 1980). The results of these experiments show that a particular species of seed is more likely to be removed if it is placed under the same, rather than another, species of tree.

Although harvester ants are often known to remove vast quantities of seeds (cf. Tevis 1958), there is conflict in the literature as to the extent to which this predation affects plant reproductive success (Reichman 1979). However, Harper et al. (1970) contend that any form of predation 'may substantially reduce the effective reproductive output of a plant' and, even though plants often produce enormous numbers of seeds, differential mortality is considered important in determining the success of plant species (Janzen 1969). There appears to be general agreement that even if ants do not seriously affect total seed supply, they may significantly influence the relative abundance of species by altering the distribution of the seeds remaining after foraging (Brown et al. 1979; Janzen 1971; Reichman 1979; Steenbergh & Lowe 1969; Tevis 1958; Whitford 1978). The results of the present study, which show high rates of removal, marked seed preferences, and considerable spatial variability in removal, indicate that harvester ants change the relative numbers and distribution of seeds on the ground and may therefore influence the distribution of plants in mallee communities. In addition, the activities of elaiosome-collecting ants may possibly increase the germination success of some species.

References

Andersen, A. N., 1980. Seed Removal by Ants at a Mallee Site in Northwestern Victoria. Unpublished Honours thesis, Monash University.

Anderson, T. W., 1960. An Introduction to Multivariate Statistical Analysis. J. Wiley & Sons, New York.

Ashton, D. H., 1979. Seed harvesting by ants in forests of *Eucalyptus regnans* F. Muell. in central Victoria. Aust. J. Ecol. 4: 265–277.

Bequaert, J., 1922. Ants in their diverse relations to the plant world. Bull. Am. Mus. Nat. Hist. 45: 333–584.

Berg, R. J., 1972. Dispersal ecology of *Vancouveria* (Berberidaceae) Am. J. Bot. 59: 109–122.

Berg, R. Y., 1975. Myrmecochorous plants in Australia and their dispersal by ants. Aust. J. Bot. 23: 475–508.

Bernstein, R. A., 1974. Seasonal food abundance and foraging activity in some desert ants. Am. Nat. 108: 490–498.

Brian, M. V., 1964. Ant distribution in a southern English heath. J. Anim. Ecol. 33: 451–461.

Briese, D. T., 1974. Ecological Studies on an Ant Community in a Semi-arid Habitat (with Emphasis on Seed-harvesting Species). Unpublished Ph. D. Thesis, Australian National University.

Briese, D. T. & Macauley, B. J., 1981. Food collection within an ant community in semi-arid Australia, with special reference to seed harvesters. Aust. J. Ecol. 6: 1–19.

Brown, E. S., 1959. Immature nutfall of coconuts in the Solomon Islands. II – Changes in ant populations, and their relation to vegetation. Bull. Ent. Res. 50: 523–558.

Brown, J. H., Reichman, O. J. & Davidson, D. W., 1979. Granivory in desert ecosystems. Ann. Rev. Ecol. Syst. 10: 201–227.

Bullock, S. H., 1974. Seed dispersal of *Dendromecon* by the seed predator *Pogonomyrmex*. Madrono 22: 378–379.

Campbell, M. H., 1966. Theft by harvesting ants of pasture seed broadcast on unploughed land. Aust. J. Exp. Agric. Anim. Husb. 6: 334–338.

Carroll, C. R. & Janzen, D. H., 1973. Ecology of foraging by ants. Ann. Rev. Ecol. Syst. 4: 231–257.

Cheal, P. D. C., Day, J. C. & Meredith, C. W., 1979. Fire in the National Parks of northwest Victoria. National Parks Service, Victoria.

Cole, A. C., 1932a. The ant, *Pogonomyrmex occidentalis* Cr., associated with plant communities. Ohio J. Sci. 32: 10–20.

Cole, A. C., 1932b. The relation of the ant, *Pogonomyrmex occidentalis* Cr., to its habitat. Ohio J. Sci. 32: 133–146.

Cole, A. C., 1934a. A brief account of aestivation and overwintering of the occident ant, *Pogonomyrmex occicentalis* Cresson, in Idaho. Can. Ent. 66: 193–198.

Cole, A. C., 1934b. An ecological study of the ants of the southern desert shrub region of the United States. Ecology 9: 388–405.

Creighton, W. S., 1966. The habits of *Pheidole ridicula* Wheeler with remarks on habit patterns in the genus *Pheidole*. Psyche 73: 1–7.

Creighton, W. S. & Creighton, M. P., 1959. The habits of *Pheidole militicida* Wheeler (Hymenoptera: Formicidae). Psyche 66: 1–12.

Culver, D. C. & Beattie, A. J., 1978. Myrmecochory in *Viola*: dynamics of seed–ant interactions in some West Virginia species. J. Ecol. 66: 53–72.

Culver, D. C. & Beattie, A. J., 1980. The fate of *Viola* seeds dispersed by ants. Am. J. Bot. 67: 710–714.

Davidson, D. W., 1977. Species diversity and community organization in desert seed-eating ants. Ecology 58: 711–724.

Greenslade, P. J. M., 1971. Interspecific competition and frequency changes among ants in Solomon Islands coconut plantations. J. Appl. Ecol. 8: 323–352.

Greenslade, P. J. M., 1973. Sampling ants with pitfall traps: digging-in effects. Insectes Soc. 20: 343–353.

Greenslade, P. J. M., 1979. A Guide to Ants of South Australia. South Australian Museum, Adelaide, 44 pp.

Greenslade, P. J. M. & Greenslade, P., 1977. Some effects of vegetation cover and disturbance on a tropical ant fauna. Insectes Soc. 24: 163–182.

Harper, J. L., Lovell, J. L. & Moore, K. G., 1970. The shapes and sizes of seeds. Ann. Rev. Ecol. Syst. 1: 327–356.

Hayashida, K., 1960. Studies on the ecological distribution of ants in Sapporo and its vicinity (1 et 2). Insectes Soc. 7: 125–162.

Hills, E. S., 1975. Physiography of Victoria: An Introduction to Geomorphology. Whitcombe and Tombs, Aust.

Hodgkinson, K. C., Harrington, G. N. & Miles, G. E., 1980. Composition, spatial and temporal variability of the soil seed pool in a *Eucalyptus populnea* shrub woodland in central New South Wales. Aust. J. Ecol. 5: 23–29.

Horvitz, C. C. & Beattie, A. J., 1980. Ant dispersal of *Calathea* (Marantaceae) seeds by carnivorous ponerines (Formicidae) in a tropical rain forest. Am. J. Bot. 67: 321–326.

Janzen, D. H., 1969. Seed-eaters versus seed size, number, toxicity and dispersal. Evolution 23: 1–27.

Janzen, D. H., 1971. Seed predation by animals. Ann. Rev. Ecol. Syst. 2: 465–492.

Johns, G. G. & Greenup, L. R., 1976. Pasture seed theft by ants in northern N.S.W. Aust. J. Exp. Agric. Anim. Husb. 16: 257–264.

Kemp, P. B., 1951. Field observations on the activity of *Pheidole*. Bull. Ent. Res. 42: 201–206.

Lawrence, C. R., 1966. Cainozoic stratigraphy and structure of the mallee region, Victoria. Proc. Roy. Soc. Vic. 79: 517–553.

Majer, J. D., 1972. The ant mosaic in Ghana cocoa farms. Bull. Ent. Res. 62: 151–160.

Majer, J. D., 1980. The influence of ants on broadcast and naturally spread seeds in rehabilitated bauxite mined areas. Reclamation Review 3: 3–9.

Majer, J. D., Portlock, C. C. & Sochacki, S. J., 1979. Ant–seed interactions in the northern jarrah forest. Abstr. Symp. Biol. Native Aust. Plants 25. Perth.

McGowan, A. A., 1969. Effect of seed-harvesting ants on the persistence of Wimmera rye-grass in pastures of North East Victoria. Aust. J. Exp. Agric. Anim. Husb. 9: 37–40.

Morton, S. R., 1982. Granivory in the Australian arid zone: diversity of harvester ants and structure of their communities. In: Barker, W. & Greenslade, J. (eds.), Evolution of the Flora and Fauna of Arid Australia. Peacock, Adelaide, in press.

Mott, J. J. & McKeon, G. M., 1977. A note on the selection of seed types by harvester ants in northern Australia. Aust. J. Ecol. 2: 231–235.

O'Dowd, D. J. & Hay, M. E., 1980. Mutualism between harvester ants and a desert ephemeral: seed escape from rodents. Ecology 61: 531–540.

Reichman, O. J., 1979. Desert granivore foraging and its impact on seed densities and distributions. Ecology 60: 1085–1092.

Russell, M. J., Coaldrake, J. E. & Sanders, A. M., 1967. Comparative effectiveness of some insecticides, repellents and seed-pelleting devices in the prevention of ant removal of pasture seeds. Trop. Grassl. 1: 153–166.

Sanders, C. J., 1970. The distribution of carpenter ant colonies in the spruce–fir forests of northwestern Ontario. Ecology 51: 865–873.

Sharpe, L. A. & Barr, W. F., 1960. Preliminary investigations of harvester ants on southern Idaho rangelands, J. Range Manag. 13: 131–134.

Shea, S. R. McCormick, J. & Portlock, C. C., 1979. The effect of fires on regeneration of leguminous species in the northern jarrah *(Eucalyptus marginata* Sm) forest of Western Australia. Aust. J. Ecol. 4: 195–205.

Smith, J. H. & Atherton, D. O., 1944. Seed harvesting and other ants in the tobacco-growing districts of North Queensland. Qld. J. Agric. Sci. 1: 33–61.

Steenbergh, W. F. & Lowe, C. H., 1969. Critical factors during the first years of life of the saguaro *(Cereus giganteus)* at Saguaro National Monument, Arizona. Ecology 50: 825–834.

Talbot, M., 1934. Distribution of ant species in the Chicago region with reference to ecological factors and physiological toleration. Ecology 15: 416–439.

Tevis, L., 1958. Interrelations between the harvester ant *Veromessor pergandei* (Mayr.) and some desert ephemerals. Ecology 39: 695–704.

Weber, N. A., 1943. The ants of the Imatong Mountains, Anglo-Egyptian Sudan, Bull. Mus. Comp. Zool. Harv. 93: 263–389.

Whitford, W. G., 1976. Foraging behavior of Chihuahuan Desert harvester ants. Am. Midl. Nat. 95: 455–458.

Whitford, W. G., 1978. Foraging in seed-harvester ants *Pogonomyrmex* spp. Ecology 59: 185–189.

Whitford, W. G. & Ettershank, G., 1975. Factors affecting foraging activity in Chihuahuan desert harvester ants. Environ. Entomol. 4: 689–696.

Willard, J. R. & Crowell, H. H., 1965. Biological activities of the harvester ant *Pogonomyrmex oweyheeii* in central Oregon. J. Econ. Entomol. 58: 484–489.

Willis, J. H., 1970. A Handbook to Plants in Victoria. Vol. I Ferns, Conifers and Monocotyledons. Melbourne University Press, Melbourne.

Willis, J. H., 1972. A handbook to Plants in Victoria. Vol. II Dicotyledons. Melbourne University Press, Melbourne.

Yasuno, M., 1963. The study of the ant population in the grassland at Mt. Hakkoda. I – The distribution and nest abundance of ants in the grassland. Ecol. Rev. Sandai 16: 83–91.

CHAPTER SIX

Ant–plant interactions in the Darling Botanical District of Western Australia

Jonathan D. Majer

Abstract. The relevant studies which have been performed in, or close to, the Darling Botanical District of the South-West Botanical Province are reviewed.

The first investigations consider the role of ants attending the flowering spike of certain *Banksia* species and the flowering parts of *Alyogyne hakeifolia*. Ants, in both cases, are considered to play a protective role against herbivores.

The interaction of ants with the mealybug *Pseudococcus macrozamiae* on the cycad *Macrozamia reidlei*, and with the scale insect *Pulvinariella mesembryanthemi* on the succulent *Carpobrotus edulis* are also discussed. Ants are considered to be of negligible importance to the survival, reproduction and colonization of the mealybug, but appear to enhance the survival and growth of scale insect populations by preventing the formation of sooty mould on honeydew.

Various aspects of ant–seed interactions in the jarrah, *Eucalyptus marginata*, forests and woodlands are described. The ant species which are elaiosome collectors or general collectors or which utilize seeds in nest construction are described. The ant–seed interactions of the two species which stand out as the most significant seed takers in the northern jarrah forest, *Rhytidoponera inornata* and *Melophorus* sp.1 (A.N.I.C.) are considered in detail. Aspects described include dietary preferences, foraging and feeding phenology, vertical distribution of seeds in ant nests, influence of nest position on seed germination patterns and also the influence of shade on ant nest distribution. Some implications of these interactions for forest ecology are discussed.

Introduction

Few investigations have been made on ant–plant interactions in Western Australia, and most examples studied have been confined to the Darling District of the South-West Botanical Province as defined by Beard (1980). This district (Fig. 1) is bounded to the west by the ocean and to the east by a line running from just north of Jurien Bay to Two People Bay near Albany. The district's climate is warm 'mediterranean' in the northern part and moderate 'mediterranean' in the south. Winter rainfall ranges from 600–1,500 mm and, over most of the district, there are 5–6 dry months per year (Beard 1980). Warren, the most southern sub-district, has a shorter dry season of only 3–4 months.

The ant–plant investigations which have been performed in or close to the Darling District include studies on ants attending flowering parts, ants attending plant-associated Homoptera and, finally, ant–seed interactions. Studies on the relationship between ants and plants have been headed by two groups, that of John Scott, formerly of the University of Western Australia, Department of Zoology (now at Entomology Branch, W.A. Department of Agriculture), and my own at the W.A.I.T. School of Biology. Some of this work is now written up although much of it is currently being prepared for publication. For convenience of identifying the projects I refer to studies in the second category by the title and author of the planned publication.

All ants mentioned in this chapter have been sort-

Buckley, R. C. (ed.), Ant-plant interactions in Australia.
© 1982, Dr W. Junk Publishers, The Hague. ISBN 90 6193 684 5.

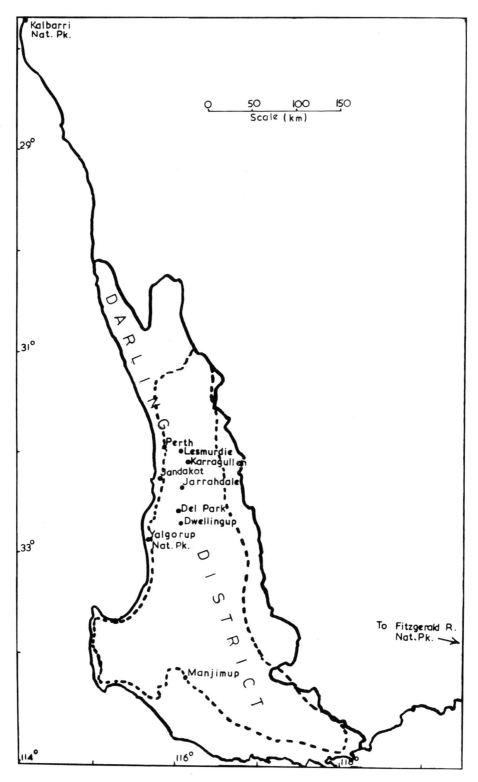

Fig. 1. Map of south-west of Western Australia showing boundary of the Darling District (solid line) and of the jarrah *(E. marginata)* forest (dotted line). The principal study sites mentioned in the text are also shown.

ed to species level. Specific names are given where possible. When unavailable they are usually either coded with Western Australian Institute of Technology (J.D.M.) code numbers or, if voucher specimens are deposited there, with Australian National Insect Collection (A.N.I.C.) codes.

Ant diversity and phenology

Since 1974, Majer (unpublished data) has been compiling a catalogue of Western Australian ants. Almost 500 species have been collected, of which approximately 300 occur within the Darling District. Although this diversity is partly accounted for by inter-site variation in species, a phenomenon associated with climatic, vegetation and other site factors, intra-site diversity is also high. For instance, grids of 6 × 6 pitfall traps with traps spaced at 3 m intervals have been established at various sites in the Darling District and have yielded between 32 and 43 species per grid (Majer 1980a; Majer & Koch, in press).

Most species show a strongly seasonal foraging activity pattern which appears to be associated principally with climatic factors. For instance, the number of individuals and species of ants collected in pitfall traps run monthly at Curara block, Dwellingup (32°51'S, 116°13'E) (Fig. 1) were low in winter, increased in spring and summer, and decreased throughout autumn (Fig. 2). Numbers of species and individuals were positively correlated with temperature and negatively with relative humidity and rainfall (Koch & Majer 1980; Majer & Koch, in press). An almost identical pattern was observed over the same study period at Perth (31°57'S, 115°47'E) (Fig. 1), although at Manjimup (34°19'S, 116°11'E) (Fig. 1), which is cooler, more humid and has a more evenly distributed rainfall, ant seasonality was rather less marked.

Pitfall traps tend to sample only the surface active species. A regular soil and litter invertebrate sampling programme performed at Karragullen (32° 04'S, 116°07'E) (Fig. 1) showed that certain hypogaeic species such as some of the Ponerinae, Dacetini and also individual species of *Oligomyrmex, Plagiolepis,* and *Stigmacros* were more active during the winter (Majer, in preparation). The remainder of this chapter will now describe some of the studies which have been involved with ant–plant interrelationships in, or close to, the Darling District.

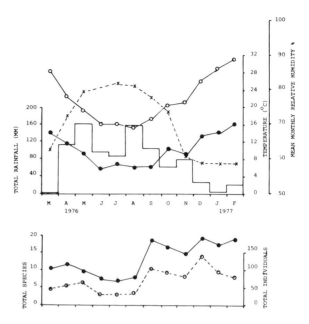

Fig. 2. Upper graph shows the monthly rainfall (histogram), mean monthly temperatures (maximum O—O, minimum ●—●), and relative humidities (x---x) at Dwellingup from March 1976 to February 1977. The lower graph shows the number of species (●—●) and individuals (O---O) of ants trapped per month at Curara Block, Dwellingup over the same period (adapted from Koch and Majer 1980; Majer and Koch, in press).

Ants attending flowering parts

Scott (1979 and in preparation) describes observations on ant attendance at flowers or associated structures of *Banksia* spp. (Proteaceae) and *Alyogyne hakeifolia* (Malvaceae). The conflorescence, or flowering spike, of certain *Banksia* is attacked by larvae of Lepidoptera and Curculionidae. Up to 90% of spikes may be attacked with a consequent reduction in seed set. In the Fitzgerald River National Park (34°00'S, 119°25'E), which is actually just outside the Darling District boundary, Scott (1979) has observed ants feeding on nectar from points resembling sites of insect damage on undeveloped *Banksia media* spikes. The liquid contained 30–51% sucrose equivalents and was taken by *Camponotus* sp., *Crematogaster* sp., *Diceratoclinea* sp., *Iridomyrmex conifer, I. purpureus* and two other species of *Iridomyrmex*. Scott (1979) suggested that the *Banksia* may produce the nectar as a response to wounding by herbivores and that the attracted ants might protect the spike from further damage. His subsequent observations supported this suggestion since seeds were

47

set along the entire conflorescence, a phenomenon which contrasts with that on other *Banksia* spp. which suffer insect damage and do not produce nectar at wound sites. An ant exclusion experiment was suggested to test this hypothesis so hopefully this question will soon be answered.

The observations on *A. hakeifolia* were also performed outside the Darling District at Kalbarri National Park (27°45′S, 114°15′E) (Scott, in preparation). Bushes were observed flowering in late November and numerous ants, including *Camponotus* sp. J.D.M. 63, *Iridomyrmex* sp. J.D.M. 9 and *I.* sp. J.D.M. 500, were present on buds, flowers, fruits and foliage of all plants observed. Nectar, containing at least 14–16% sucrose equivalents, was produced at the joint of the bracts. No distinct nectary organ was found and it was suggested that hairs in these regions might have a secretory function. Although ants attending buds were aggressive to intruders, Scott's measurements showed that 88% of flowers exhibited signs of herbivore damage. He suggested that presence of ants might reduce the level of herbivore activity at fruits and flowers. Experiments to test this hypothesis were recommended.

Ant–Homoptera–plant interactions

Two ant–Homoptera–plant interactions have been investigated in *Banksia* woodland near Jandakot (32°10′S, 115°50′E). The first examined the survival and reproduction of the mealybug, *Pseudococcus macrozamiae* (Pseudococcidae) on *Macrozamia reidlei* (Cycadaceae) in burnt and unburnt areas and on plants where ants had been excluded (Dolva & Scott, in press). The mealybug, which is species specific in the study area, favours protected sites on the host such as the bases of leaves, where protection is afforded by a cotton-like growth from the bulb, and on new growth where the leaflets are held together to form a partially enclosed environment. Fire stimulates mealybug build-up since it encourages new growth, and the resulting 'crinkling' of damaged leaves and adhesion of leaflets by resin from wounds also produce new enclosed habitats. Although fire destroys mealybugs on exposed leaves, those at the base of leaves may survive. The mealybug is tended by *Iridomyrmex chasei*, *Camponotus* sp. and to a lesser extent, other species. Dolva and Scott measured the numbers and lengths of leaves, the presence of crinkled habitats and the numbers of mealybugs in areas burnt 5 months, 2 years and at least 10 years previously. Plants which had not been burnt for at least 10 years had fewer leaves, fewer distortions to leaflets and fewer mealybugs than the more recently burnt areas. Having established that fire stimulates mealybug habitat formation, and also mealybug abundance, they then investigated how important ants were for survival and colonization of the insect. Mealybug clusters on plants in the area burnt 2 years previously were isolated from their attendant ants by banding the leaf bases with 'Tanglefoot'; other clusters were designated as controls. In addition, artificial mealybug habitats were constructed by taping leaflets together on plants in the area burnt 5 months previously, where few clusters were yet present. The build-up of mealybug numbers was then followed for 115 days in artificial habitats where ants were either present or artificially excluded. Ants were not found to have any relationship with mealybug survival or colonization in either experiment, there being no difference between experiment and control in mealybug cluster size and number of young in the first experiment, or mealybug colonization in the latter. It was, however, established that larger mealybug clusters were attended by larger numbers of ants, presumably owing to greater honeydew availability. The study concluded that although fire stimulates new growth and provides new mealybug habitats, the ant–mealybug relationship is facultative and is apparently of negligible importance to survival, reproduction and colonization by the mealybug.

The other study performed at Jandakot revealed a scale insect–ant relationship in which ants were of major importance to the homopteran (Collins & Scott, in preparation). Here the relationship between the scale insect *Pulvinariella mesembryanthemi* (Coccidae) and the prostrate succulent *Carpobrotus edulis* (Aizoaceae) was investigated. This plant is a native of South Africa and is now naturalized in many coastal areas where an endemic member of the genus also exists. The scale is probably host specific and is also an introduction. It is eaten by adults and larvae of *Cryptolaemus montrousieri* (Coccinellidae) and is attended by the ants *Crematogaster* sp. J.D.M. 33, *Iridomyrmex conifer* and *I.* sp. J.D.M. 9.

The investigators asked two questions: firstly, do the coccinellids limit the number of scales, and secondly, do the ants benefit the scales by removing honeydew or by reducing predation? To answer these questions they devised three experimental treatments. Ants, but not predators, were excluded from one set of branches by grease banding the edges of cloth placed between the plant and the ground. Predators, but not ants, were excluded from another set by constructing tents of 2 mm net over branches. Both treatments were combined on a third set of branches, and a fourth set was left as a control. Scales were counted 6 times over a period of 18 days. The experiment revealed that predators had only a minor limiting effect on the scales. No reduction in predation could be detected where ants were present. Although more mortality was observed where ants were excluded, the ant + predator exclusion treatment produced very similar results. The investigators concluded that the importance of ants to the scale is not that of reducing predation but of preventing the growth of sooty mould by the removal of honeydew from the scales. The relationship is facultative since some of the scales were not tended by ants. However, presence of ants was considered to be necessary for the formation of large populations of scales.

Ant–seed interactions

Introduction

A considerable body of information has now been gathered on seed removal by ants in the jarrah *(Eucalyptus marginata* Myrtaceae) forest and woodland of Western Australia. The thrust for such work has come from two sources. First is the Alcoa bauxite mining venture which, after mining is completed, rehabilitates the land by a range of methods. Investigations have been carried out to see if ants might impede direct seeding of the mined area by taking seeds (Majer 1978, 1980b). Second is the suggestion that leguminous understorey plants such as *Acacia* spp. might be encouraged in the jarrah forest because they suppress the pathogenic soil fungus, *Phytophthora cinnamomi,* the causal agent of jarrah die-back disease (Anon. 1978). Since seeds of such plants possess elaiosomes and are frequently collected by ants (Berg 1975), the role of ants in the dispersal and survival of legume seeds has been investigated in detail (Shea, McCormick & Portlock 1979; Majer, Portlock & Sochacki 1979; Majer & Portlock, in preparation). In what follows I extract and present the ecological findings of these studies but do not

discuss the application of the findings to the two applied problems mentioned.

Census of seed-harvesting ants in the northern jarrah forest

The principal seed-harvesting ants in the northern jarrah forest were censused by a number of techniques. Between 1975 and 1979 records were kept of any species observed carrying seeds. The middens of rejected food and other items, which surround the nest entrances of many species, were examined for seeds, and nests were also excavated to search for seed stores. Finally, seed depots of jarrah (*E. marginata)* and *Acacia extensa,* a myrmecochorous species, were established at various sites and the ants observed removing seeds were noted (Majer 1980b). Where possible, species were recorded as elaiosome collectors, general collectors or species which utilized seeds in nest construction (decorators), following the terminology of Berg (1975). Nine species were found to take seeds consistently, as part of their diet or for nest construction (Table 1). It was not possible to give a rigid classification of collecting behaviour – *Rhytidoponera inornata,* for instance, usually ate only elaiosomes, though it consumed some seeds almost entirely – but most species showed a tendency towards one particular type of behaviour. Most commonly occurring sub-families contained at least two representative seed takers. With the exception of *Camponotus* sp., all genera have previously been recorded taking seeds in Australia; Berg (1975) recorded this genus as a non-collector. Overall, taking ant abundance and seed content of diet into account, two species stand out as the most significant seed takers in the northern jarrah forest; *Rhytidoponera inornata* and *Melophorus* sp. 1 (A.N.I.C.).

Rates of seed removal by ants

At least three factors are likely to affect seed attractiveness to ants: the presence of an elaiosome, its size and its chemical attractiveness. Majer (1980b) compared the rate of removal of seed in jarrah forest at Jarrahdale (32° 18′S, 116° 05′E) and Del Park, near Dwellingup (32° 40′S, 116° 02′E). Masonite boards 12 cm square were installed throughout the forest at each site with their rough side uppermost. Ten seeds each of *A. extensa* (mean fresh weight 11.5 mg/seed) and *E. marginata* mean fresh weight 14.0 mg/seed)

Table 1. Seed-taking ants in the northern jarrah forest, and uses to which seeds are put.

Sub-family	Species	Feeds on elaiosome only	Feeds on entire seed	Uses seed for nest structuring
Ponerinae	*Rhytidoponera inornata*	usually	occasionally	–
	Rhytidoponera violacea	–	usually	–
Myrmicinae	*Pheidole latigena*	–	usually	–
	Meranoplus sp.12 (A.N.I.C.)	usually	occasionally	–
Formicinae	*Melophorus* sp.1 (A.N.I.C.)	usually	occasionally	–
	Camponotus sp. J.D.M. 199	?	?	?
Dolichoderinae	*Iridomyrmex purpureus*	–	–	occasionally
	Iridomyrmex conifer	?	?	?
	Iridomyrmex sp. 19 (A.N.I.C.)	?	?	?

were placed on each board in the morning of each observation period. Only the *Acacia* seeds possessed elaiosomes. Depots were then inspected 7 times during the following 24 hours to count the number of seeds which had been removed and make observations on seed taking. The Jarrahdale and Del Park experiments were performed in November 1977 and December 1977 respectively. Figure 3 shows the mean cumulative number of seeds removed per depot for each species at each site. Overall, 48–66% of *A. extensa* but only 3–9% of *E. marginata* were removed within 24 hours. There were no significant differences between the two sites, and the overall pattern and level of seed removal was surprisingly similar, with seeds of *A. extensa* 7.5 times more likely to be removed, on average, than those of *E. marginata*. Ants observed taking seeds from depots included *Rhytidoponera violacea, Camponotus* sp. 15 (A.N.I.C.), *Melophorus* sp. 1 and *Iridomyrmex* sp. 19 (A.N.I.C.).

These seed-removal rates may have been inflated by presenting seeds in clumps rather than a more diffuse pattern, but they are nevertheless similar to those observed for pasture seeds by Russell, Coaldrake & Sanders (1967), Johns & Greenup (1976) and Mott & McKeon (1977). The most important result is that *A. extensa* seeds were removed in preference to those of *E. marginata*. Since the two species have similar seed weights, the difference in removal rates seems likely to be due to the *Acacia* elaiosomes. This is supported by previous records of elaiosome collection (Berg 1975) for at least two of the ants involved, namely *Melophorus* sp. 1 and *Iridomyrmex* sp. 19 (A.N.I.C.). Preliminary experiments have also shown that considerably fewer *Acacia* seeds are taken by ants if the elaiosomes are experimentally

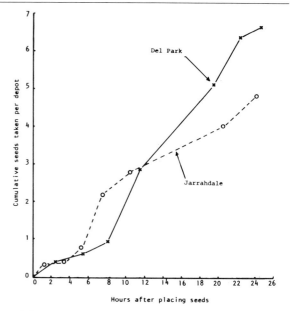

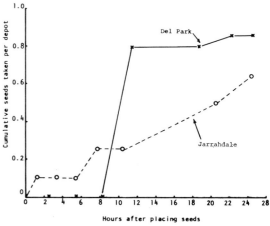

Fig. 3. Mean cumulative number of seeds removed by ants from depots established in forest at Jarrahdale and Del Park. The upper graph shows data on *Acacia extensa,* the lower on *Eucalyptus marginata* (jarrah) (adapted from Majer 1980b).

50

removed (Majer, unpublished). The alternative possibility remains, however, that the *Acacia* seeds are chemically more attractive to the ants. Ashton (1979), for instance, has suggested that certain sugar-like substances on the surface of *Eucalyptus regnans* seeds may be involved in the attractiveness of seed to ants. Further experiments are currently being performed to separate the effects of these two possible influences.

The influence of seed size has recently been investigated using three species of *Eucalyptus: E. calophylla* (mean fresh weight 119.0 mg/seed), *E. marginata* (14.0 mg/seed) and *E. wandoo* (0.4 mg/seed). Forty boards were installed in forest at Lesmurdie (32°01′S, 116°03′E) in April 1980, and 24 hours of observations were made along the same lines as the previous experiment. Observations were made from 0900 hrs onwards for the two species with larger seeds. The lighter *E. wandoo* seeds were placed out at 1400 hrs and observed for 12 hours only, since the remainder of the day was too windy to carry out observations. Figure 4 shows the mean cumulative number of seeds removed per depot for the observation period. Although the *E. wandoo* values may be due partly to seed being blown away by wind, the data suggest that larger non-elaiosome bearing seeds are less likely to be removed by ants than are small ones. Alan Andersen (this volume) has observed the same phenomenon when rates of seed removal of

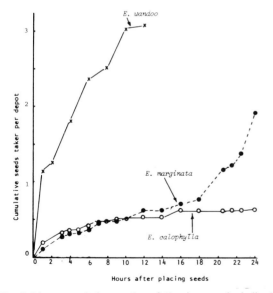

Fig. 4. Mean cumulative number of *Eucalyptus calophylla, E. marginata* and *E. wandoo* seeds removed by ants from depots established in woodland at Lesmurdie (Majer, unpublished data).

different *Eucalyptus* spp. were compared in a north-western Victoria mallee region. The Lesmurdie experiment also indicated a possible relationship between seed weight and ant head capsule width. *Rhytidoponera violacea* (mean head capsule width 1.7 mm) took *E. calophylla* seeds; *Melophorus* sp. 1 and *Pheidole latigena* (mean head capsule widths 0.7 and 0.6 mm) took *E. marginata;* and *Iridomyrmex* sp. indet. (mean head capsule width 0.6 mm) took *E. wandoo.* More observations are needed, however, to establish the generality of this relationship.

Influence of elaiosome removal on seed viability

The various surveys on seeds collected by jarrah forest ants have yielded a large checklist of species taken by ants. Seeds are either stored within the nest or rejected in the middens around nest entrances. If the ant is a general collector (Berg 1975), then all that generally remains of the seed on the midden is part of its outer coat. In the case of elaiosome-collecting ants, the seed is usually placed on the midden after the elaiosome has been removed.

Majer & Portlock (in preparation) examined the seeds rejected by one elaiosome collector, *Melophorus* sp. 1. Plates 1a and b show the elaiosomes of two commonly collected seeds; *Bossiaea aquifolium* (Leguminosae) and *Acacia strigosa.* Plates 1c and d show parts of seeds of the same genera which were collected from the middens of *Melophorus* sp. 1. In both cases the elaiosome has been removed at the hilum leaving the testa intact. Germination tests were performed on seed collected from middens around five nests of *Melophorus* sp. 1: after immersion in boiling water, they were placed on filter paper over moist vermiculite in petri dishes at 25 °C. Seventy-two percent of the seeds were found to be viable (Table 2), a value equivalent to that obtained for seeds of these species collected directly from the plant.

Table 2. Viability of *Acacia strigosa* and *Phyllanthus calycinus* seed from *Melophorus* sp.1 nest middens.

| | Nest number | | | | | |
	1	2	3	4	5	$\overline{\overline{X}}$
Total seeds	27	9	18	98	10	32.4
Total viable	23	3	13	82	9	26
Percent viable	85	30	72	84	90	72

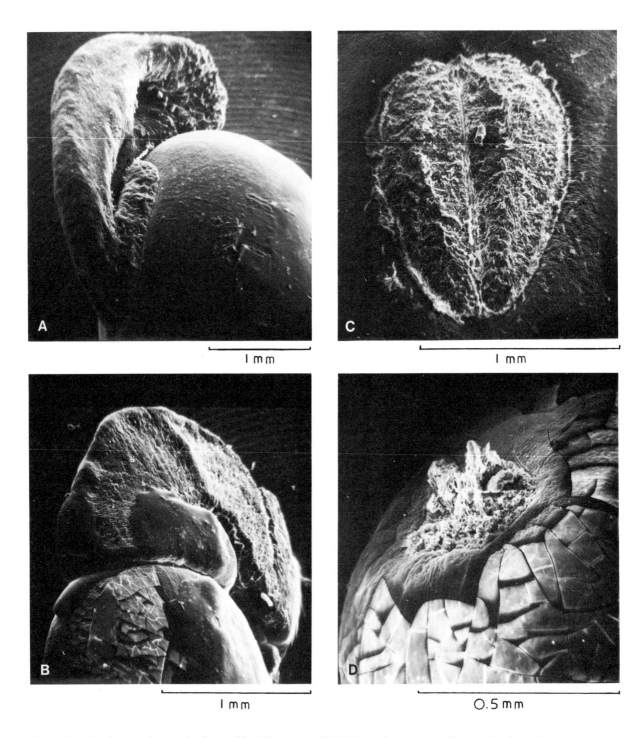

Plate 1. Scanning electron micrographs of parts of fresh *Bossiaea aquifolia* (A) and *Acacia strigosa* (B) seeds showing elaiosome and also the areas where this structure has been removed by *Melophorus* sp. 1 from *B. aquifolium* (C) and *Acacia* sp. (D) seeds.

Comparison of Rhytidoponera inornata and Melophorus sp. 1

As mentioned earlier, *R. inornata* and *Melophorus* sp. 1 are considered to be the most significant seed takers in the northern jarrah forest. Both species are elaiosome collectors although they sometimes consume entire seeds (Table 1). The following discussion summarizes a study carried out to compare the ecology and seed-taking activities of these two species (Majer & Portlock, in preparation).

Dietary preferences were assessed by two techniques. The first involved collection and identifica-tion of rejected food fragments from middens around nests of each species, by floating off organic matter from the soil and sorting and identifying individual food units under a stereo-microscope. Secondly, at Karragullen, five nests of both species were each observed for 30-minute periods at monthly intervals between March 1978 and April 1979, and all food items being carried by workers entering the nests were collected for later identification. Unfortunately, their low foraging density and timid nature resulted in the failure of this method for *R. inornata*, though it was successful for the other species. Table 3 shows the midden content analyses. Both species are omni-

Table 3. Composition of nest middens at Karragullen. *R. inornata* middens collected February 1978, *Melophorus* sp.1 middens collected November 1978.

(a) *Melophorus* sp. 1	Nest number										
	1	2	3	4	5	6	7	8	9	10	Mean
Arthropod fragments	15	1	4	67	10	33	–	5	1	28	16.4
Seeds											
Eucalyptus marginata (Myrtaceae)			1	6	4					2	1.3
Acacia strigosa (Leguminoseae)					1					1	0.2
Phyllanthus calycinus (Euphorbiaceae)						1				3	0.4
Trymalium ledifolium (Rhamnaceae)	60					2					6.2
Miscellaneous spp.	13			6	5	33		13	5	3	7.8
% of all items	49% seeds, 51% arthropods										

(b) *Rhytidoponera inornata*	Nest number					
	1	2	3	4	5	Mean
Arthropod fragments	72	241	24	78	461	175.2
Seeds						
Eucalyptus marginata (Myrtaceae)	263	3	5	18	27	75.2
Eucalyptus calophylla (Myrtaceae)	3	–	–	18	–	4.2
Casuarina sp. (Casuarinaceae)	–	–	1	3	–	0.8
Gompholobium tomentosum (Leguminoseae)	1	3	2	–	–	1.2
Acacia pulchella (Leguminoseae)	–	2	–	–	–	0.4
Acacia sp. indet. (Leguminoseae)	1	1	27	–	–	5.8
Miscellaneous spp.	4	37	42	7	–	18.0
% of all items	37% seeds, 63% arthropods					

vorous, with arthropod fragments in the middens as well as seeds. More food fragments were found around *R. inornata* nests than around those of the other species, suggesting that this species consumes more food per colony. Arthropod fragment counts overestimate consumption, since one such food item may be broken into many fragments; the seeds, in contrast, were all intact apart from the elaiosome. If total fragments are used as an index of food preference, Table 3 shows that *Melophorus* sp. 1 relies more on seeds than *R. inornata*. The year's cumulative feeding records show that the diet of *Melophorus* sp. 1 comprises 37% seeds, 24% miscellaneous plant fragments and 39% invertebrates. The few successful field-feeding observations made on *R. inornata* confirmed the midden analysis finding that this species is

less reliant upon seeds and more on invertebrates.

The seasonal activity patterns of these two ants are related to their feeding habits. Grids of pitfall traps run regularly through 1975–6 at Dwellingup and 1978–9 at Karragullen produced the catches shown in Figs. 5a and 5b for *R. inornata* and *Melophorus* sp. 1 respectively. *R. inornata* foraged throughout the year, albeit with a winter trough, whereas *Melophorus* sp. 1 exhibited a summer peak and ceased foraging completely from March or May until October. Majer (1981) censused the plants species flowering each fortnight in and around the Karragullen ant survey plot. The total numbers of species in flower during each observation period are shown in Fig. 5c. Flowering was low throughout most of the year but rose to a peak in October and then declined

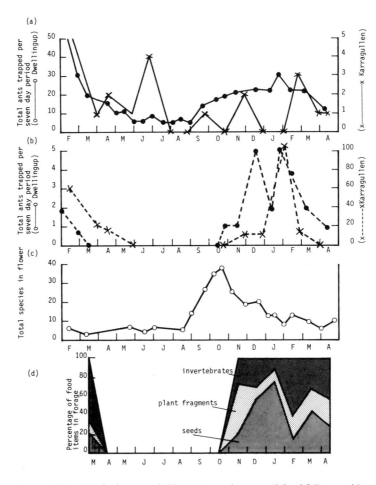

Fig. 5. Total *Rhytidoponera inornata* (a) and *Melophorus* sp. 1 (b) ants trapped per month in pitfall trap grids run at Dwellingup between 1975–76, and Karragullen between 1978–79. The third graph (c) shows the total number of plant species in flower at the Karragullen site between 1978–79, and (d) indicates the percentage composition of food items collected from foraging *Melophorus* sp. 1 workers at Karragullen over the same period (from Majer and Portlock, in preparation).

to its former level. Seed production was not measured directly, but most species would produce seed in the period of 1 to 3 months after flowering. The monthly composition of the *Melophorus* sp. 1 diet, measured as above, is shown in Fig. 5d. The proportion of seeds is high in January, three months after flowering, after which it declines and is replaced by invertebrate and other plant material. In addition (though not shown in the figure), total food consumption per colony per half hour observation period was much greater in January than in any other month. Greenslade (1979) has pointed out that *Melophorus* is a sun-loving genus which generally forages when conditions are hot. The results presented here suggest that warm weather foraging may be an adaptation to coincide with high available seed levels. Observations by Majer & Portlock (in preparation) show that aboveground seed stocks are depleted by the onset of winter, so that the cessation of foraging in winter may be due simply to the lack of suitable food. Although there are no data on monthly variations in diet composition for *R. inornata,* its year-round foraging probably reflects the ability to switch from a high-seed diet during summer to a winter diet rich in invertebrates, particularly those associated with decomposition on the forest floor (Koch & Majer 1980). There are also differences in diurnal foraging patterns: *Melophorus* sp. 1 forages only during the day, when seeds are most likely to be dropped, whereas *R. inornata* is nocturnal during the warmer months and diurnal only during winter.

Considerable data are now available on the vertical distribution of seeds within *R. inornata* and *Melophorus* sp. 1 nests. In sites near Dwellingup, Shea et al. (1979) found that seed was much more abundant in ant nests than in the soil, and that seed was stored primarily at depths of 0–3 cm, with smaller numbers down to 12 cm. Majer and Portlock (in preparation) have studied seasonal variations in the patterns of seed storage by *Melophorus* sp. 1, and have also investigated the relationship between seed profile and nest depth for both species. In both studies 400 cm³ cores of soil were taken from nests of each species. Cores were 3 cm thick and were taken from four depths: 0–3, 3–6, 6–9 and 9–12 cm. No seeds were found below 12 cm, so no cores were taken below this depth. Similar sets of samples were taken from soil randomly located 1 m from each nest. For *Melophorus* sp. 1, 10 samples were taken in May 1977, 6 in April 1977, 6 in November 1977 and 5

in January 1978. The first two dates were sampled by, and reported in, Shea et al. (1979). Four *R. inornata* nests were sampled in February and May 1977 by Shea et al. (1977). The mean seed numbers at each depth for *R. inornata* and *Melophorus* sp. 1 are shown in Fig. 6 and 7 respectively. Sample dates are rearranged to show the effects of season, and only legume seeds are included in the counts (principally *Acacia* spp., *Bossiaea aquifolium* and *Mirbelia dilatata).* The legume seed count ranged from 20 to 82 seeds per nest for *Melophorus* sp. 1, and from 17 to 163 for *R. inornata.* Seed levels in the soil controls ranged from 1 to 6 per sample. Seasonal trends in the seed profiles agree with those found by Shea et al. (1979), save that for the January records of *Melophorus* sp. 1 nests, more seed was found at the 6–12 cm level than at 0–6 cm. January is the main period of seed fall so seed may initially be stored in the lower chambers of nests. The seed density in nests of *Melophorus* sp. 1 was low in November and doubled or quadrupled following seed fall (Fig. 7). The data on *R. inornata* were inadequate to compare with this trend, but the drop in nest seed levels between February and May (Fig. 6) indicated that this species did not store seed for long periods. This contrasts with the behaviour of *Melophorus* sp. 1, which had a high nest seed count in April (Fig. 7), just prior to cessation of aboveground foraging (Fig. 5). Nest depth was also examined directly for both species (Majer & Portlock, in preparation), by pouring molten lead into nests (Ettershank 1968) and excavating the

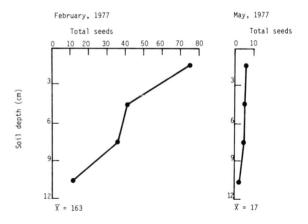

Fig. 6. Occurrence of legume seeds at various depths in *Rhytidoponera inornata* nests situated near Dwellingup. Counts for February 1977 and May 1977 are shown separately and the mean number of legume seeds per nest are indicated (adapted from Shea et al. 1979).

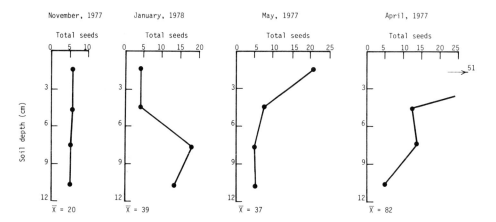

Fig. 7. Occurrence of legume seeds at various depths in *Melophorus* sp. 1 nests situated near Dwellingup. Data for the November pre- and three post-seed fall periods are shown separately and the mean number of legume seeds per nest are indicated (November and January data are from Majer and Portlock, in preparation; the remainder are adapted from Shea et al. 1979).

cast. Mean nest depths for *Melophorus* sp. 1 and *R. inornata* in the laterite soils of the Darling Range were 12 ± 3 cm and 23 ± 11 cm respectively, and most nest chambers of *R. inornata* were at depths of less than 12 cm. Nest depths therefore tally closely with seed depth profiles for both species.

In sandy soils on the coastal plain at Yalgorup National Park (115°32'E, 32°40'S), *Melophorus* sp. 1 nests are considerably deeper than in the lateritic soils, ranging from 20 ± 4 cm in a low-lying area to 24 ± 10 cm on an upland soil (Majer & Portlock, in preparation). The differences between the two Yalgorup sites may be related to levels of soil moisture or to the water table level. The mean seed profile values for the low-lying area ($n = 10$) and the upland ($n = 3$) are shown in Fig. 8. The April 1977 Dwellingup values are also shown for comparison, as are traces of typical nests from all three localities. Nest depth and seed profile are least for lateritic soils, higher in low-lying sandy soils, and greatest in upland sandy soils.

Seed-germination patterns are also influenced by the distribution of ant nests. Shea et al. (1979) noted that legumes tended to germinate in clumps following hot fires and postulated that seeds were germinating on the sites of ant nests. Subsequently Majer & Portlock (in preparation) selected 22 *R. inornata* nests at Hollyoake, near Dwellingup, and 15 and 10 nests respectively of *Melophorus* sp. 1 and *R. inornata* at Marradong Road, near Dwellingup. The former site was in an area of high plant species diversity, the latter in a plantation of exotic *Eucalyp-*

tus sp. with a low diversity understorey of native plants. Four treatments were used: in the first, a 700 cm² area of soil over each nest was heated in autumn with a 'Pyrox Schwank' infra-red heater to reach 100 °C at 2 cm for 30 minutes; the second comprised similar heating of bare soil areas 1 m from each nest; the third, unheated nests; and the fourth, unheated soil areas. Germinated seedlings were counted and identified in midwinter.

The Hollyoake data for *R. inornata* and the Marradong Road data for both species are shown in Tables 4 and 5 respectively. Eight seedling species were common. Germination was consistently higher from nest areas than bare soil. Heating increased germination in species such as *Acacia pulchella* and *Trymalium ledifolium* (Fig. 5), but reduced rates for others such as *Eucalyptus marginata*, *Acacia urophylla*, *Bossiaea ornata*, *Mirbelia dilatata* and *Gompholobium marginatum* (Table 4). Excavations at the Marradong Road site showed that most seedlings on the nests of both ants had germinated from seeds at 1.4–2.1 cm depth. This corresponds well with actual seed burial depths in nests (Figs. 6 and 7); stimulation of germination without seed incineration will be critically dependent on the precise soil temperature profile. The distribution of seedlings around three representative heated *R. inornata* nest entrances at Hollyoake is shown in Fig. 9. Seedlings are most abundant on the nest or midden. For nests on a slope, species such as *E. marginata* germinated in a 'tear-drop' pattern, the tip representing seeds which have rolled or washed down slope from the midden

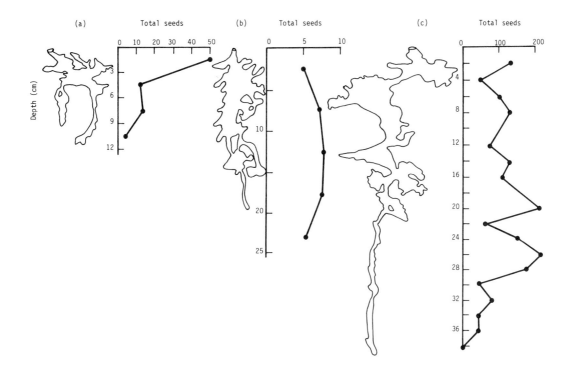

Fig. 8. Occurrence of seeds at various depths in *Melophorus* sp. 1 nests situated near Dwellingup (a), and in Yalgorup National Park (low-lying area (b) and upland (c)). The Dwellingup data are for legume seeds only, Yalgorup counts are for all seeds. Diagrams of typical *Melophorus* sp. 1 nests from the three areas are also shown and are drawn to the same scale as the graphs (from Majer & Portlock, in preparation).

Table 4. Seedling frequencies in July 1978 in heated and unheated 700 cm² areas with or without *R. inornata* nests, at Hollyoake (site 3). Em = *Eucalyptus marginata,* As = *Acacia* sp., Au = Acacia urophylla, Bo = *Bossiaea ornata,* Md = *Mirbelia dilatata,* Gm = *Gompholobium marginatum.*

	Heated nest						Heated soil					
	Em	As	Au	Bo	Md	Gm	Em	As	Au	Bo	Md	Gm
Mean number of seedlings	1.7	0.7	0	5.1	1.4	0.2	0	0.1	0	0	0.1	0.1
S.d.	4.8	1.9	0	21.2	5.5	0.6	0	0	0	0	0.3	0
Frequency	4	4	0	6	4	2	0	1	0	0	3	1
Max. seedlings per site	21	8	0	100	26	2	0	1	0	0	1	3

	Unheated nest						Unheated soil					
	Em	As	Au	Bo	Md	Gm	Em	As	Au	Bo	Md	Gm
Mean number of seedlings	6.1	0	0.1	13.6	4.7	0.4	0.1	0	0	0.1	0.2	0
S.d.	11.3	0		7.4	5.5	0	0	0	0	0	0.5	0
Frequency	2	0	1	5	3	1	1	0	0	1	3	0
Max. seedlings per site	29	0	1	25	11	3	1	0	0	1	2	0

Number of observations: 22.

57

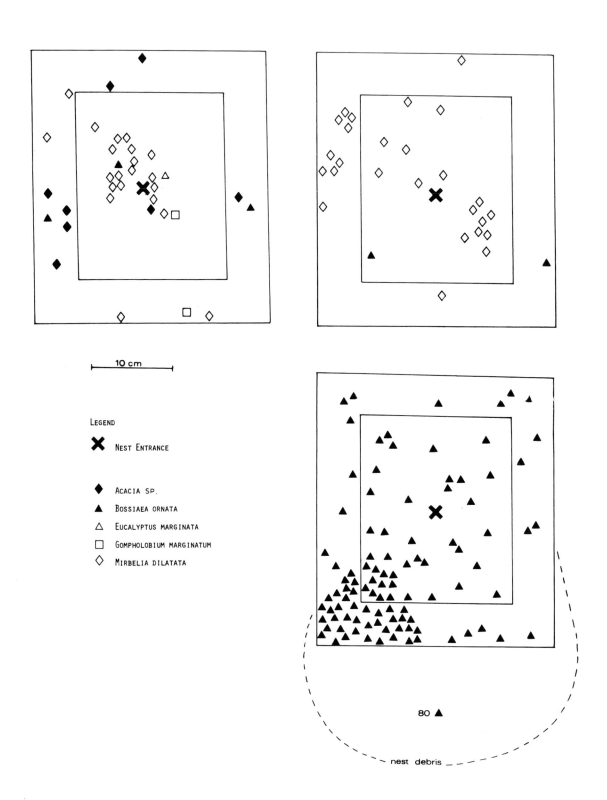

Fig. 9. Maps of seedlings which have germinated near the entrances of three *Rhytidoponera inornata* nests which were heated up to 100 °C at 2 cm depth. The inner rectangle represents the area of the heating apparatus, the outer rectangle incorporates soil which was heated by lateral conduction from the heat source (from Majer & Portlock, in preparation).

Table 5. Seedling frequencies in July 1979 in heated and unheated 700 cm² areas with or without ant nests, at Marradong Road (site 2). Ap = *Acacia pulchella*, Tl = *Trymalium ledifolium*.

	Heated nest		Heated soil		Unheated soil		Unheated nest	
	Ap	Tl	Ap	Tl	Ap	Tl	Ap	Tl
(a) *Melophorus* sp.1								
Number of observations	15	15	15	15	18	18	3	3
Mean number of seedlings	0.5	2.6	0.3	0.5	0	0	0	0
S.d.	1.0	3.7	0.5	0.8	0	0	0	0
Frequency	4	9	5	4	0	0	0	0
Max. seedlings per site	3	14	1	2	0	0	0	0
(b) *Rhytidoponera inornata*								
Number of observations	10	10	10	10	10		3	3
Mean number of seedlings	0.3	8.9	0.1	1.1	0	0.1	0	0
S.d.	0.7	17.0	0	2.8	0	0	0	0
Frequency	2	5	1	3	0	1	0	0
Max. seedlings per site	1	27	1	9	0	1	0	0

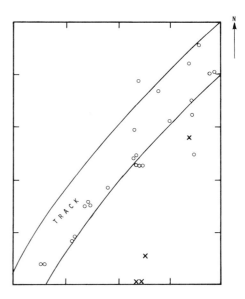

Fig. 10. Map of *Rhytidoponera inornata* (x) and *Melophorus* sp. 1 (0) nests in a 20 × 25 m area of forest at Karragullen (from Majer and Portlock, in preparation).

Nest distribution of *R. inornata* and *Melophorus* sp. 1 may be influenced by shade. The nest distributions of both species were mapped in a 20 × 20 m plot at Marradong Road in December 1977, and percent ground cover and topshade estimated visually within each 1 m square of the plot. *Melophorus* sp. 1 nests were also mapped in a 72 × 130 m plot in sandplain at Yalgorup National Park, and canopy density estimated in each 3 m plot square. Actual nest occurrences were then compared with expected occurrences estimated from the relative frequency of each shade class. Chi-square analyses showed no significant effect of shade on nest position for *R. inornata*, but significantly greater frequencies of *Melophorus* sp. 1 nests in less shaded areas at both sites. This was confirmed by mapping the nests of *Melophorus* sp. 1 and *R. inornata* in an evenly shaded 20 × 25 m plot at Karragullen, cut by an abandoned forestry track. The nests were mapped in March 1978 (Fig. 10); four *R. inornata* nests were situated in vegetated areas, but the 25 *Melophorus* sp. 1 nests were almost all on the track, and the remainder at dead trunk bases under sparse ground cover.

Discussion

The studies outlined above locally substantiate Berg's (1975) observation that seed taking is a widespread phenomenon in Australia. The seeds of at least 29 jarrah forest plant species, from 21 genera, are taken by elaiosome-collecting, decorator or seed-harvesting ants. Of these, elaiosome collectors are most prominent (Table 1 and unpublished ant abundance data). This is in marked contrast to more arid areas; Briese & Macauley (1977), for example, found that seed-harvesting *Chelaner* and *Pheidole* spp. constituted 42% of the ant colonies in a semi-arid *Atriplex vesicaria* plain in New South Wales. Few *Pheidole* spp. are found in the northern jarrah forest, and colonies of *Chelaner* spp. are small and localized. Two hypotheses may be advanced for this paucity of seed harvesters. The first is that seed types used by harvesters are uncommon in the northern jarrah forest; it could be tested by quantifying seed output and suitability to harvesters. The second is that though a flush of seeds is produced following the October flowering peak, uncollected seed on the surface may deteriorate under moist winter conditions until valueless to ants, in contrast to arid environments where seeds are probably available to ants all year. Seeds are a major dietary item for semi-arid *Pheidole* and *Chelaner* spp. throughout the year (Briese 1974), but only at seed fall for *Melophorus* sp. 1 (Fig. 5d). Further tests require comparison of ground seed loads and decomposition rates in jarrah forest and arid areas.

One major result of the jarrah forest studies is to indicate the magnitude of seed removal by elaiosome collectors. Nest densities of *Melophorus* sp. 1 and *R. inornata* approximate 1350 nests per ha (Majer & Portlock, in preparation); in view of the nest seed content values given above this is equivalent to 123,000 legume seeds per ha. Bare ground seed density is lower, but represents 4.5 million seeds per ha overall since the area is much larger than that occupied by ants. Hence only 2.7% of total legume seeds are apparently in ant nests. Similarly, 4.7% of seedlings are located on ant nests at Marradong Road (Table 5); the higher value may be due to the inclusion of the non-legume *Trymalium ledifolium*. An alternative approach, involving direct observation of *Acacia* seed at seed fall (Majer & Portlock, in preparation) suggests that the above estimates of percentage seed held in ant nests are gross underestimates. In addition, experiments using seed depots (Majer 1980b) indicate that 48–66% *A. extensa* seeds may be removed in 24 hours (Fig. 3). These may be overestimates, but measured values for seed contents of nests underestimate total harvest because of seed rejection and seed loss in abandoned galleries or old nests. Hence the true levels of harvesting probably lie between the lower and upper estimates.

Since *Melophorus* sp. 1 nests preferentially in open areas and ant nests act as foci for seedling germination, clearings should eventually become vegetated by plants originating from the nests. This is supported by causal observations indicating that cleared areas do become colonized by myrmecochorous plants; a mechanism which would restore breaks in ground cover or canopy.

The widespread occurrence of myrmecochory in Australian plants (Berg 1975) may have evolved in response to needs for additional dispersal agents, or for protection, by burial, from bushfires, predation and the diurnal microclimatic cycle. The jarrah forest experiments cannot separate the roles of these influences, but it is noteworthy that most seed in ant nests is at an ideal depth for post-fire seed germination. There are no significant differences in total nitrogen or total or available phosphorus contents between soils from *R. inornata* and *Melophorus* sp. 1 nests and soils from control areas 1 m from each nest (10 nests of each species, Majer, unpublished) so enhancement of seedling nutrient status by ant excreta or rejecta is unlikely to be a contributing factor.

In conclusion, there is now a substantial body of information on ant–seed interactions in the northern jarrah forest. Some results, however, require additional experiments for confirmation and for comparison with arid areas. The relative roles of ants and mammals (Morton 1982) also requires investigation.

References

Anon, 1978. A destructive fungus in Australian forests. Ecos 15: 3–14.

Ashton, D. H., 1979. Seed harvesting by ants in forests of *Eucalyptus regnans* F. Muell, in central Victoria. Aust. J. Ecol. 4: 265–277.

Beard, J. S., 1980. A new phytogeographic map of Western Australia. West. Aust. Herb. Res. Notes 3: 37–58.

Berg, R. Y., 1975. Myrmecochorous plants in Australia and their dispersal by ants. Aust. J. Bot. 23: 475–508.

Briese, D. T., 1974. Ecological Studies on an Ant Community in a Semi-arid Habitat. Ph. D. Thesis, Australian National University.

Briese, D. T. & Macauley, B. J., 1977. Physical structure of an ant community in semi-arid Australia. Aust. J. Ecol. 2: 107–120.

Collins, L. & Scott, J. K., in preparation. Interaction of ants, predators and the scale insect *Pulvinariella mesembryanthemi* on *Carpobrotus edulis*, an exotic naturalised in Western Australia.

Dolva, J. M. & Scott, J. K., 1982. The association between the mealybug, *Pseudococcus macrozamiae* (Homoptera), ants and the cycad, *Macrozamia reidlei* in a fire prone environment. J. Roy. Soc. West. Aust., in press.

Ettershank, G., 1968. The three dimensional gallery structure of the nest of the meat ant *Iridomyrmex purpureus* (Sm.) Hymenoptera: Formicidae). Aust. J. Zool. 16: 715–723.

Greenslade, P. J. M., 1979. A Guide to the Ants of South Australia. South Australian Museum Special Educ. Bull. Series, Adelaide, 44 pp.

Johns, G. G. & Greenup, L. R., 1976. Predictions of likely theft by ants of oversown seed for the northern tablelands of New South Wales. Aust. J. Exp. Agric. Anim. Husb. 16: 257–264.

Koch, L. E. & Majer, J. D., 1980. A phenological investigation of various invertebrates in forest and woodland areas in the south-west of Western Australia. J. Roy. Soc. West. Aust. 63: 21–28.

Majer, J. D., 1978. Studies on invertebrates in relation to bauxite mining activities in the Darling Range. A review of the first eighteen months research. Alcoa Environ. Res. Bull. 3: 18.

Majer, J. D., 1980a. Report on a study of invertebrates in relation to the Kojonup nature reserve fire management plant. West. Aust. Inst. Tech. Biol. Dep. Bull. 2: 22.

Majer, J. D., 1980b. The influence of ants on broadcast and naturally spread seeds in rehabilitated bauxite mined areas. Reclamation Review 3: 3–9.

Majer, J. D., 1981. A flowering calendar for Karragullen, a northern jarrah forest locality. West. Aust. Herb. Res. Notes 5: 19–28.

Majer, J. D., in preparation. The seasonal dynamics of an ant community at Karragullen, a northern jarrah forest locality.

Majer, J. D. & Koch, L. E., 1982. Seasonal activity of hexapods in woodland and forest leaf litter in the south-west of Western Australia. J. Roy. Soc. West. Aust., in press.

Majer, J. D. & Portlock, C. C., in preparation. A comparison of the seed-taking activities of two Western Australian ants, *Melophorus* sp. 1 (A.N.I.C.) and *Rhytidoponera inornata*.

Majer, J. D., Portlock, C. C. & Sochacki, S. J., 1979. Ant–seed interactions in the northern jarrah forest. Abstr. Symp. Biol. Native Aust. Plants 25. Perth.

Morton, S. R., 1982. Granivory in the Australian arid zone: diversity of harvester ants and structure of their communities. In: Barker, W. & Greenslade, J. (eds.), Evolution of the Flora and Fauna of Arid Australia. Peacock, Adelaide, in press.

Mott, J. J. & McKeon, G. M., 1977. A note on the selection of seed types by harvester ants in northern Australia. Aust. J. Ecol. 2: 231–235.

Russell, M. J., Coaldrake, J. E. & Sanders, A. M., 1967. Comparative effectiveness of some insecticides, repellents and seed-pelleting devices in the prevention of ant removal of pasture seeds. Trop. Grassl. 1: 153–166.

Scott, J. K., 1979. Ants protecting *Banksia* flowers form destructive insects? West. Aust. Nat. 14: 151–154.

Scott, J. K., 1981. Extrafloral nectaries in *Alyogyne hakeifolia* (Giord.) Alef, (Malvaceae) and their association with ants. J. Roy, Soc. West. Aust, 15: 13–15.

Shea, S. R., McCormick, J. & Portlock, C. C., 1979. The effect of fires on regeneration of leguminous species in the northern jarrah *(Eucalyptus marginata* Sm.) forest of Western Australia. Aust. J. Ecol. 4: 195–205.

CHAPTER SEVEN

Ant-epiphytes of Australia

Camilla R. Huxley

Abstract. Symbioses between ants and plants are widespread in the tropics. Trees are usually defended by their ants, while epiphytic plants probably obtain mineral nutrients via the ants. In northern Australia three Malesian genera of ant-epiphytes are found: *Myrmecodia* and *Hydnophytum* (Rubiaceae) and *Dischidia* (Asclepiadaceae). *Myrmecodia* and *Hydnophytum* form large tubers with cavities, some of which have specialized warts, probably for absorption. In *Dischidia* the leaves form sacs which are entered by adventitious roots. Ants, usually *Iridomyrmex cordatus,* occupy these cavities and sacs. They deposit organic remains in the cavities with warts. Radio-isotopes fed to the ants are later absorbed by the plants. The behaviour of the ants and other occupants of the ant-plant are discussed in relation to the symbiosis.

The three species of *Myrmecodia* and two of *Hydnophytum* in Australia require moist tropical conditions, but are absent from the rain forest. In coastal and savanna vegetation, where other epiphytes are rare, they are abundant, probably due to improved mineral nutrition from the symbiosis. In *Dischidia* transpirational water loss may be reduced by the symbiosis, enabling the plants to use drier habitats.

Illustrations and a key to Australian species are given together with nomenclatural notes. Species of ants which have been found in ant-plants are tabulated.

Introduction

In tropical areas throughout the world there are plants which produce hollow structures in which ants regularly nest. Many of these plants are trees for which the ants play a defensive role, either by killing insect pests, attacking mammalian herbivores or chewing at plants which would compete with the host tree. The ants usually inhabit hollow thorns, internodes, or leaf-pouches, and they often obtain some form of food from the tree: either oil- or protein-rich food bodies or extra-floral nectar. Symbioses of this type are not known in Australia, with the possible exception of *Ficus subinflata* Warb. which Bailey (1909) says has hollow internodes inhabited by ants.

There are a number of epiphytic plants in Australia which harbour ant nests in specific chambers, but the role of the ants appears to be nutritive rather than defensive. The ants bring food and waste matter into some of the chambers of the plants. Breakdown products of this material are absorbed and used by the plants. Such a source of nitrogen and phosphorus compounds is of particular advantage to epiphytes, which do not have access to the soil.

In tropical America this habit is adopted only by a few species in one bromeliad genus and one small genus of ferns. In the Far East and Australia, however, over 150 species in three plant families are involved in this kind of symbiosis with ants. In Australia the best-known ant-epiphytes are two genera in the Rubiaceae: the ant-plants *Hydnophytum* and *Myrmecodia* (Figs. 1–4). These plants with their large, rounded tubers form a conspicuous element in the mangrove and coastal swamps around Cairns and in the forests of northern Cape York. Less well-known are the asclepiad *Dischidia rafflesiana* (Fig. 5) and the fern *Lecanopteris sinuosum,* not found in Australia (Huxley 1980).

Buckley, R. C. (ed.), Ant-plant interactions in Australia.
© 1982, Dr W. Junk Publishers, The Hague. ISBN 90 6193 684 5.

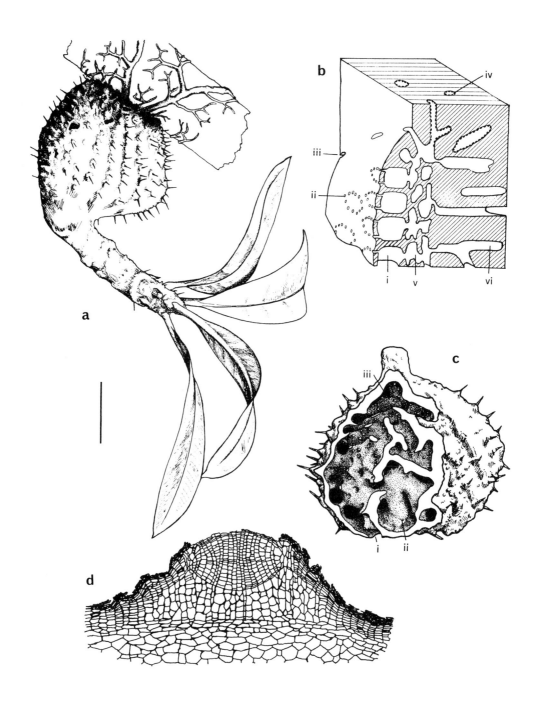

Fig. 1. *Myrmecodia tuberosa.* a. Whole plant; scale = 5 cm. b. Block diagram of part of tuber of a variant of *M. tuberosa* from Manus Island, Papua New Guinea; scale = 1.3 cm. i. one 'cell' of honeycomb. ii. pore in ring at top of a cell. iii. entrance hole at periphery of honeycombed area. iv. warted tunnel. v. smooth, dark tunnel in open network area. vi. smooth, shelf-like, straw-coloured tunnel. c. Tuber of young plant cut away to show the third cavity; scale = 1.4 cm. i. entrance hole. ii. inner smooth chamber. iii. dark brown, warted tunnels. d. Longitudinal section of a wart; scale = 0.7 mm. [After Huxley (1980), reproduced with permission from *Biological Reviews.*]

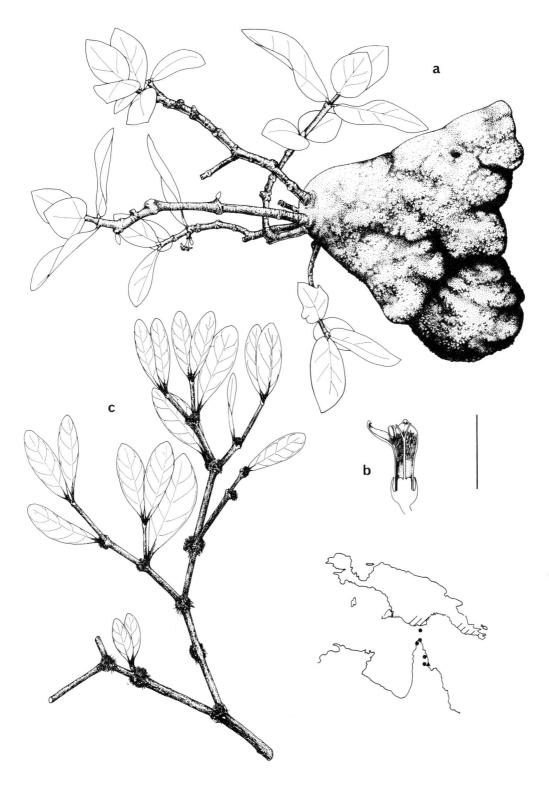

Fig. 2. Hydnophytum. a. *H. papuanum* whole plant; scale = 8 cm. b. Flower of *H. papuanum;* scale = 0.4 cm. c. *Hydnophytum* sp. 1; scale = 4.8 cm. On map ● = localities of *H. papuanum;* ▲ = localities of H. sp. 1. Hatching = probable extent of *H. papuanum* in New Guinea.

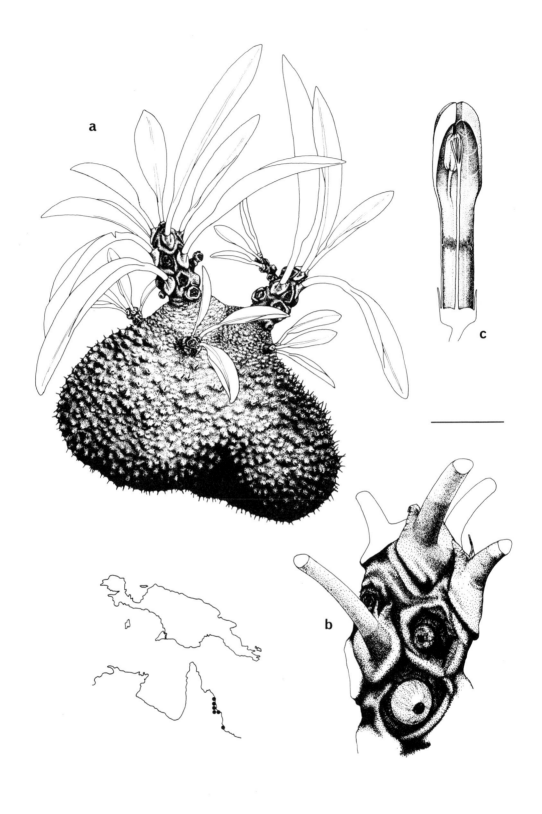

Fig. 3. Myrmecodia beccarii. a. Whole plant; scale = 6 cm. b. Stem detail; scale = 1 cm. c. Flower; scale 0.4 cm.

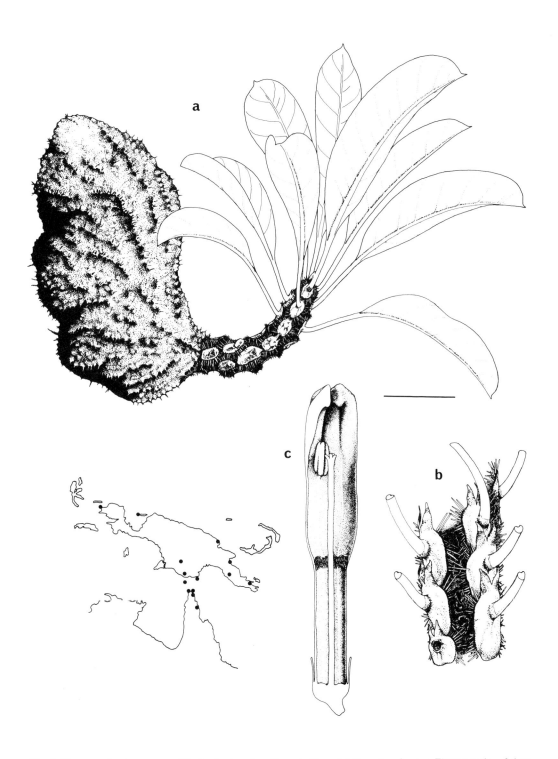

Fig. 4. Myrmecodia platytyrea. a. Whole plant; scale = 8 cm. b. Stem detail; scale = 2 cm. c. Flower; scale = 0.4 cm.

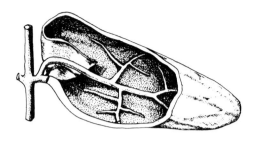

Fig. 5. Dischidia rafflesiana. Leaf cut open to show roots; scale = 2 cm. [After Huxley (1980), reproduced with permission from *Biological Reviews.*]

Strikingly, a single ant species, *Iridomyrmex cordatus,* inhabits these plants throughout the drier lowland parts of their range. A second species, *I.* cf, *scrutator,* replaces *I. cordatus* in the wetter and the mid- to high-altitude regions of Papua New Guinea. Other ants occur sporadically and segregates of *I. cordatus* are recognized in some areas.

Structure and physiology of the plants

The ant-plants Myrmecodia and Hydnophytum

In *Hydnophytum* and *Myrmecodia* the formation of the tuber and its chambers occurs irrespective of the presence of ants (Forbes 1885; Treub 1883, 1888). As the young seedling grows the hypocotyl swells, and when it is about 1 cm across the first cavity is formed by a layer of impermeable cells. The tissue within this layer dies and dries out. Ants probably aid in the process of removing the epidermis to form an entrance from the outside. More cavities are created by successive impermeable layers. These later cavities are usually more complex in shape and are also differentiated into chambers with different types of surface (Fig. 1b, c).

The surfaces of the cavities vary in both *Hydnophytum* and *Myrmecodia;* they may be pale or dark brown, and they may or may not be studded with little white swellings referred to as warts (Fig. 1d). In *Hydnophytum* the warts are usually most dense at the tips of blind tunnels. In *Myrmecodia* the structure is more complex, the smooth chambers often

forming distinct shelf-like areas, while the warted chambers are tunnel-shaped. There may also be smooth honey-combed areas just under the surface of the tuber which are much darker brown than the smooth, shelf-like areas (Fig. 1b, c). The tunnels of some *Myrmecodia* tubers are remarkably similar to the arrangement of tunnels in some termite mounds (Matthew Jebb, personal communication).

The anatomy of the warts was first described by Treub (1883) who thought they were involved in gas exchange. Later Miehe (1911) showed that the areas with warts absorb water rapidly while the chambers without warts do not absorb water. He also found that stains applied to the warted surfaces are readily taken up by the cells of the warts but not by the general surface. He therefore suggested that the function of the warts is to absorb nutrients in solution from the cavities.

Recent studies with radio-isotopes have shown that substances fed to the ants in honey or in fly larvae are absorbed and translocated by the plants (Huxley 1978; Rickson 1978). The ants deposit phosphorus-32 in the warted chambers rather than the smooth ones, presumably by defaecation (Huxley 1978). Rickson suggests that breakdown products of food stored but not eaten by the ants are used by the plants.

Other ant-epiphytes

Rather less is known about the structure and function of the nesting sites in *Dischidia* and *Lecanopteris.* In *Dischidia* the ants nest in sacs formed by the leaves. Most members of the genus have biconvex leaves, for instance *D. nummularia* R.Br., a common creeping epiphyte in northern Queensland. In some members of the genus, not found in Australia, the leaves are strongly concave downwards with the edges of the leaves appressed to the host tree: ants nest in the resulting cavities. In other species, notably *D. rafflesiana,* some of the leaves are modified to complete sacs (Fig. 5). Adventitious roots arising from near the petiole enter the sacs. The ants appear to use the sacs initially for their larvae and pupae. Later they fill the sacs with debris and the adventitious roots then proliferate throughout the debris. Besides nutrients from this source it appears that the plants also absorb carbon dioxide from within the sacs, since the inner (abaxial) but not the outer, surface is richly supplied with stomata. In *Lecanop-*

teris species the chambers occupied by ants are formed by the disintegration of the central tissue of the rhizomes (Jermy and Walker 1975).

Ant occupants

Species involved

A study of the literature (Table 1) and my own records from Papua New Guinea (Fig. 6) indicate that *Iridomyrmex cordatus* (Fr. Smith) and closely related species or subspecies are the usual ants in

Table 1. Records of ants in ant-plants.

Ant species	Ant-plant	Authority
Ancyridris sp. nr. *polyrachioides*	*Hydnophytum*	P. Ward, p.c.
Camponotus sp.	*Myrmecodia* various	Mann, 1921 B.B. Lowery, p.c.
Camponotus maculatus	*M. dahli*	Dahl, 1901
Camponotus maculatus	*M. tuberosa*	Miehe, B1922*
Camponotus quadrisectus	*M.* sp.	Wheeler, B1922
Crematogaster sp.	*M. alata*	Beccari, B1922
Crematogaster sp.	*M. erinacea*	Beccari, B1922
Crematogaster difformis	*M. tuberosa*	Beccari, B1922
	H. montanum	Beccari, 1884
	M. tuberosa	Bequaert, 1922
	M. sp	Wheeler, B1922
Iridomyrmex cordatus	*M. goramensis*	Beccari, B1922
	M. bullosa	Beccari, B1922
	M. albertisii	Beccari, B1922
	H. petiolatum	Beccari, B1922
	H. montanum	Miehe, B1922
	M. pentasperma	Dahl, 1901
Iridomyrmex myrmecodiae	*M. tuberosa*	Beccari, B1922
	M. guppyanum	Mann, 1919
	M. tuberosa	Miehe, 1911
	M. tuberosa	Shelford, B1922
	M. antoinii	Bequaert, 1922
Iridomyrmex nagasau	*Myrmecodia*	Mann, 1921
Paratrechina sp.	*M.* or *H.* sp.	B.B. Lowery, p.c.
Pheidole sp.	*M. & H.* sp.	Mann, 1921
	M. or *H.* sp.	B.B. Lowery, p.c.
Pheidole javana	*M. & H.* spp.	Wheeler, 1926
Pheidole megacephala	*Myrmecodia*	Barrett, 1928
	M. rumphii	Beccari, B1922
	M.alata	Beccari, B1922
Pheidole myrmecodiae	*M.* or *H.* sp.	Montieth, 1974
Poecilomyrma senirewae	*M. & H.* spp.	Mann, 1921
Solenopsis sp.	*M. & H.* spp.	B.B. Lowery, p.c.
Strumigenys sp.	*M.* or *H.* sp.	B.B. Lowery, p.c.
Technomyrmex albipes	*H.* sp.	P. Ward, p.c.
Tetramorium pacificum	*M. & H.* spp.	B.B. Lowery, p.c.

* B1922 = In Bequaert (1922).

Myrmecodia and *Hydnophytum* through the drier lowland part of their range. Records from *Lecanopteris* and *Dischidia* show that these are also inhabited by *I. cordatus*. Some authors have distinguished forms of *I. cordatus* living in ant-epiphytes as *I. myrmecodiae* Emery, but Dr R. W. Taylor does not feel that it is possible to subdivide *I. cordatus* satisfactorily. *I. nagasau*, reported by Mann (1921) to occupy *Hydnophytum* species in Fiji, is also related to *I. cordatus*. Observations from Papua New Guinea show that in the rain forest and high-altitude forests, *I. cordatus* is replaced by members of another large and variable species of *Iridomyrmex*, which is referred to by Dr R. W. Taylor as *I.* cf. *scrutator*. Ant-plants have not to my knowledge been collected from the rain forest of Queensland and therefore not from habitats where one would expect this ant to be present. Other ants are recorded quite frequently from ant-plants (Table 1) but none of them on a regular basis. My observations in Papua New Guinea suggested that a variety of ants may invade ant-plants in disturbed areas, especially close to habitation. *Pheidole* may occupy a significant proportion of Australian ant-plants; *P. megacephala* is mentioned by Barrett (1928) and *P. myrmecodiae* by Monteith (1974).

Ant behaviour

Individual colonies of *I. cordatus* usually occupy many ant-plants and perhaps several host trees (Janzen 1974). It seems that the same colony may use cavities of all the ant-epiphytes on the tree. The ants usually construct carton covering over their numerous runways. Various vegetable matter including seeds is incorporated into this carton so that one often sees rows of seedling ant-plants along the runways. The build-up of carton around the roots is probably beneficial for the plants.

Iridomyrmex cordatus appears to be omnivorous. In Borneo and on the Malaya Peninsula this ant feeds on honey-dew from scale insects, though this was not common in Papua New Guinea (Huxley 1978). These ants certainly feed on other insects and have been seen removing moribund individuals. The presence of numerous heads of their own species inside the ant-plant cavities suggests that they may eat members of the neighbouring colony. They probably also eat oil-bodies on the seeds of *Dischidia rafflesiana* and nectar from the flower discs of ant-

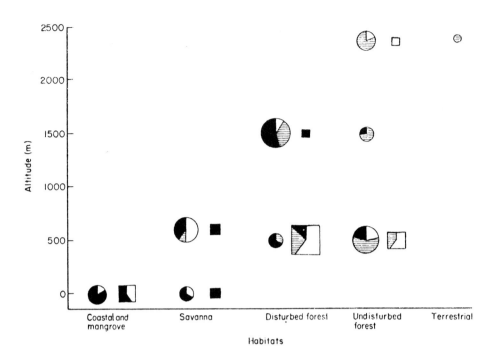

Fig. 6. Distribution of ants in ant-plants from different habitats. Circles represent *Myrmecodia,* squares represent *Hydnophytum.* The area of each symbol is proportional to the number of records. Each symbol is divided so that the angles represent the relative frequency of *Iridomyrmex cordatus* (shaded areas), *I.*cf.*scrutator* (hatched areas), and other ant species (white areas). [After Huxley (1978), reproduced with permission from *New Phytologist.*]

plants after the corollas have fallen, and on either or both of the fungi found in the cavities (see next section).

The success of the symbiosis with the ant-epiphytes depends however on the behaviour of the ants within the cavities. Observation of the distribution of ant juveniles and debris in the tubers of ant-plants shows that the juveniles are generally kept in the wart-free, smooth-walled areas, while the debris is in the warted tunnels, especially in their distal portions. If the ants are fed with radioactive honey, considerable radioactivity appears on the warted walls but not the smooth ones.

Other organisms inhabiting ant-plants

Fungi

During the work in Papua New Guinea two kinds of fungus were regularly found in ant-plant tubers, one on the smooth-walled chambers and one in the warted ones. Both of these fungi were found in my

collections of *M. beccarii* from Cairns (UPNG 3493 and UPNG 3494). The fungus on smooth walls appeared to be parasitic, the mycelium being within the *Myrmecodia* cells, and only the conidia projecting into the cavities. Ants may eat the spores, since no spores were seen on the conidia. In the warted chambers another monilialine fungus *(Arthrocladium)* is found. It seems to be saprophytic on the organic matter in the cavities and again it is possible, but unconfirmed, that the ants feed on it (Miehe 1911; Huxley 1978). Various other organisms are also found in the decaying debris; for instance dipteran larvae, nematodes and protozoa. The activity of these organisms must result in many compounds being solubilized and thus made available to the plants.

Lycaenids and other Lepidoptera.

Larvae of the Lycaenid butterfly *Hypochrysops apollo* Miskin feed on the leaves and tubers of *Myrmecodia* species in Queensland. The variety *H. apollo apollo* is found on *M. beccarii* near Cairns,

70

and *H. apollo phoebus* on *Myrmecodia* sp. on Cape York (Waterhouse 1928; Common and Waterhouse 1972). The larvae are normally found with *Iridomyrmex cordatus,* though it is not clear whether they are tended by the ants. In some areas the introduced *Pheidole megacephala* inhabits *Myrmecodia* and appears indifferent to the larvae. Another species, *H. arronica,* has been reared from *Myrmecodia* in southern Papua New Guinea. These plants were associated with an *Iridomyrmex* species, probably *I.* cf. *scrutator.* Larvae of a leaf-miner moth, *Cangetta aurantiaca* (Pyralidae), are abundant in *Myrmecodia* leaves in southern Papua, New Guinea (Huxley 1978). These are the main insect pests of *Myrmecodia,* probably since the larvae are protected from the ants by the leaf epidermis. I have not found records of leaf-miners in Australian ant-plants.

Distribution, ecology and conclusion

All the ant-epiphytes in Australia form part of the Malesian element in the flora of northern Queensland. Their overall distributions extend from Burma and Indochina through Malaysia, the Philippines, Indonesia and Papua New Guinea to Fiji. In Queensland, *Hydnophytum* and *Myrmecodia* are confined to the less arid vegetation of the extreme north, and the east coast as far south as Ingham (Fig. 7). They are absent from pure *Eucalyptus* savanna, since the peeling bark deters epiphytes; the few plants found on *Eucalyptus* were perched in the forks of large branches. Ant-plants in Australia are most commonly found on species of *Melaleuca* (often on white sands). They are also abundant on members of the Myrtaceae in white sand areas of both Borneo and the north of the Malaya Peninsula. The white quartz sand is a highly leached and mineral-poor soil. In both these areas *Iridomyrmex cordatus* appears to be feeding largely on honey-dew from scale insects on the trees (Janzen 1974). Unlike the other species, *Myrmecodia beccarii* is usually found in mangrove and coastal swamps; e.g. *Rhizophora* in the Cairns inlet. The small fleshy leaves of this species and the high proportion of tissue to cavity in its tubers probably represent adaptations to a dry and sometimes salty environment. Unfortunately nothing is known of any differences in habitat preference between *Hydnophytum papuanum* and the two species of *Myrmecodia* on the Cape York Peninsula.

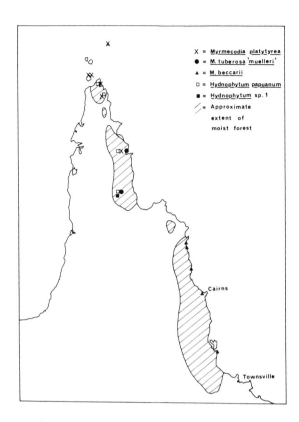

Fig. 7. Map of distribution of ant-plants in Australia compared with the distribution of moist forest.

It is undoubtedly the result of the symbiosis with ants that ant-plants are common in the open canopies of swamps and savannas where other epiphytes are rare, and rare in rain forest, where there are many other epiphytes. In closed canopies there is a much greater accumulation of dead leaves, moss and ferns, which provides reserves of the nutrients and water necessary for epiphytes. In open canopies such accumulation is rare, and the supply of nutrients via ants, and the protection of seedlings by the ant runways, may be crucial to the survival of epiphytes. The ant-epiphyte which is most resistant to exposure to the sun, and indeed grows readily on dead trees, is *Dischidia rafflesiana* (Fig. 5). It is notable that this plant has stomata on the inside of the pitchers and is therefore able to absorb carbon dioxide produced by the ants or by their decomposing debris. This would save the plant from the water loss which is normally entailed by opening the stomata to obtain carbon dioxide from the air. Open canopies, moreover, are much harder for ants to colonize as the scarcity of dead branches and accumulated litter reduces the number of nesting sites available. Thus, as one might expect, the symbiosis is only common in environments which either symbiont would find very difficult to occupy without the other.

Knowledge of these plants and their relationship with ants and Lycaenids is still fragmentary, and further studies will doubtless reveal many unsuspected facts. As an aid to field observations I include a key and illustrations here (a monograph of *Myrmecodia* is in preparation). I would like to thank Eleanor Huxley for the illustrations.

Key to Australian ant-plants

1. Tuber without spines; stems woody, usually less than 0.8 cm across, internodes usually more than 1.5 cm long; leaves sessile or short stalked .... *Hydnophytum*. . . 2
1. Tuber more or less spiny; stem fleshy, usually more than 0.8 cm across, internodes less than 1.5 cm long; leaves stalked ..... *Myrmecodia* 3
2. Flowers borne on mounds in the leaf axils, surrounded by a dense mass of reddish brown hairs; leaves coriaceous, elliptical .... *Hydnophytum* sp. 1
2. Flowers not surrounded by a mass of reddish brown hairs; leaves fleshy, ovate .... *Hydnophytum papuanum*
3. Leaves fleshy, usually less than 12 cm long; petiole green, c. 2 cm long; stems several and freely branched; fruit white...... *Myrmecodia beccarii*
3. Leaves thin or leathery, usually larger; petiole reddish or white, longer than 2 cm; stem solitary and unbranched or with only a few branches; fruits red or orange 4
4. Stems covered by spiny, shield-like outgrowths at the base of each leaf, petiole often reddish....... .... *Myrmecodia platytyrea*
4. Stem without shield-like outgrowths but with elongated depressions more or less filled by bracts and hairs; petiole white or green .... *Myrmecodia tuberosa* 'muelleri'

Nomenclature of Australian ant-epiphytes

1. *Hydnophytum papuanum* Becc.
The plants usually referred to as *H. formicarum* in Australia are probably not conspecific with the plants from Sumatra which were given this name by William Jack (1823). The Australian plants closely resemble specimens from New Guinea which were described by Beccari as *H. papuanum* (1884–6). Further work, however, is necessary to confirm this identification: in the meantime, I will refer to the Australian plants as *H. papuanum* (Fig. 2a, b).

2. *Hydnophytum* sp.1
There is one collection of *Hydnophytum* from Australia which undoubtedly belongs to a different species. This was collected by Brass at Leo Creek in the McIlwraith Range (Brass 1885). Rather similar plants have been found in northwest and southeast New Guinea. Further work is needed to determine whether these belong to the same species and to describe them (Fig. 2c).

3. *Myrmecodia platytyrea* Beccari
Beccari (1884–6) described a species *M. antoinii* on the basis of a collection from Thursday Island which F. von Mueller had previously referred to as *M. echinata*. Beccari admitted that this species is very similar to *M. platytyrea* Becc. from Papua New Guinea. *M. antoinii* has shorter shield-like structures along the stem, smaller leaves and relatively shorter petioles. Taxonomic work now in progress shows that these two taxa are not distinct and should both be called *M. platytyrea* Becc. (Fig. 4).

4. *Myrmecodia beccarii* Hook f.
First discovered by Joseph Banks, this plant was forgotten till Joseph Hooker (1886) described it from a specimen sent to Kew. It was reported to have come from the Gulf of Carpentaria, but this seems unlikely in view of its present known distribution on the east coast of Queensland from Cooktown to Ingham. This is the only species of ant-plant which is confined to Australia (Fig. 3).

5. *Myrmecodia tuberosa* 'muelleri'
The name *M. muelleri* was first applied to collections from the Fly River in southern Papua New Guinea. The Australian plants are very similar, and should be placed in the same taxon. These plants form a southern continuation of *M. tuberosa* Jack, a variable species distributed throughout Malaysia and Indonesia. (F. von Mueller also used the name *M. echinata*). In the forthcoming taxonomic treatment of *Myrmecodia* they will be referred to an informal

subdivision of *M. tuberosa* called *M. tuberosa* '*muelleri*'.

6. *Dischidia* species

Bailey (1900) records the presence of *D. nummularia* R.Br. with leaves up to 1 cm across, and *D. ovata* Benth. with leaves over 2 cm long. These two species lack leaf sacs. Bailey (1900) also mentions *D. timorensis* Dcne. and *D. rafflesiana* Wall., and Schlechter later described *D. baeuerlenii* from Thursday Island. These all possess leaf sacs, and probably all belong to one species, which is best called *D. rafflesiana* R.Br. for the time being (Fig. 5).

Little is known of the distribution and requirements of these species and their symbionts.

References

Bailey, F. M., 1900. The Queensland Flora, 773–775. Diddams, Brisbane.

Bailey, F. M., 1909. Comprehensive Catalogue of Queensland Plants. Queensland Government, Brisbane.

Barrett, C., 1928. Ant-house plants and their tenants. Victorian Naturalist, 133–137.

Beaglehole, J. C., 1962. The Endeavour Journal of Joseph Banks 1768–1771. Trustees of the Public Library of New South Wales, in association with Angus and Robertson, Sydney.

Beccari, O., 1884–6. Genoa: Tipografia del istitudo Sordo-Muti. Malesia, 2.

Bequaert, J., 1922. Ants and their diverse relations to the plant world. Bull. Am. Mus. Nat. Hist. 14: 522–528.

Cochrane, G. R., 1963. A physiognomic vegetation map of Australia. J. Ecol. 51: 639–655.

Common, F. I. B. & Waterhouse, D. F., 1972. Butterflies of Australia. Angus and Robertson, Sydney.

Dahl, F., 1901. Das Leben der Ameisen im Bismarck-Archipel. Mitt. zool. Mus., Berlin 2: 1–69.

Forbes, H. O., 1885. A Naturalist's Wanderings in the Eastern Archipelago. Sampson, Low, Marston, Searle and Rivington, London.

Huxley, C. R., 1978. The ant-plants *Myrmecodia* and *Hydnophytum* (Rubiaceae) and the relationships between their morphology, ant occupants, physiology and ecology. New Phytol. 80: 231–268.

Huxley, C. R., 1980. Symbiosis between ants and epiphytes. Biol. Rev. 55: 321–340.

Jack, W., 1823. Account of the *Lansium* and some other genera of Malayan plants. Linn. Soc. Lond., Trans. 14: 122–124.

Janzen, D. H., 1974. Epiphytic myrmecophytes in Sarawak, Indonesia: mutualism through the feeding of plants by ants. Biotropica 6: 237–259.

Jermy, A. C. & Walker, T. G., 1975. *Lecanopteris spinosa* – a new ant-fern from Indonesia. Fern Gazette 11: 165–176.

Mann, W. M., 1919. The ants of the British Solomon Islands. Bull. Mus. Comp. Zool. Harv. 63: 362–406.

Mann, W. M., 1921. The ants of the Fiji Islands. Bull. Mus. Comp. Zool. Harv. 69: 469–472.

Miehe, H., 1911. Über die javanische *Myrmecodia* und die Beziehung zu ihren Ameisen. Biol. Zbl. 31: 733–737.

Monteith, G. B., 1974. Focus on Cape York. Rep. Fauna Subcomm. Entomol. Soc. Queensl., September 19th.

Rickson, F. R., 1979. Absorption of animal tissue breakdown products into a plant stem: the feeding of a plant by ants. Am. J. Bot. 66: 87–90.

Treub, M., 1883. Sur le *Myrmecodia echinata* Gaudich. Ann. Jard. Bot. Buitenz. 3: 129–159.

Treub, M., 1888. Nouvelles reherches sur le *Myrmecodia* de Java. Ann. Jard. Bot. Buitenz. 7: 191–212.

Valeton, T., 1927. Die Rubiaceae von Papuasien. Bot. Jahrb. 61: 127–150.

Walter, H. 1964. Vegetation der Erde, 452. Gustav Fischer, Jena.

Waterhouse, G. A., 1928. Notes on Australian Lycaenidae, Pt.4. Proc. Linn. Soc. N.S.W. 53: 405–407.

CHAPTER EIGHT

Plants' use of ants for dispersal at West Head, New South Wales

Mark Westoby, Barbara Rice, Julia M. Shelley, David Haig and J. L. Kohen

Abstract. Of the world's known species of myrmecochores, plants which provide food bodies to induce ants to disperse their seeds, many are found in the dry sclerophyll vegetation of Australia. Here we present data and observations on myrmecochory on the West Head, and area of dry sclerophyll vegetation near Sydney, and in the light of these we discuss possible explanations for the distribution of myrmecochory. In most stands myrmecochores made up about 30% of the overall plant species complement, but few of the dominant species were myrmecochores. Myrmecochores tended to occur in a wider range of stands than non-myrmecochores. Within a stand, they tended to occur more of their own diameters from their nearest conspecific neighbours, but otherwise did not occupy detectably different microsites. Many proposed explanations for myrmecochory could explain either its commonness in Australia, or why it should be adaptive in sclerophyll shrubs, but none explain why it should be adaptive in sclerophyll shrubs in Australia but not in California or the Mediterranean. Australian dry sclerophyll vegetation, unlike that of California or the Mediterranean, is delimited by low-phosphorus soils; empirically, the correlation of myrmecochory with low-phosphorus soils is good. Evidence is given to suggest that in the vegetation of the West Head the limiting currency for seed production is phosphorus. Food bodies cost little phosphorus, so that myrmecochory is cheap in terms of the effective currency. However, wings and hairs are also cheap in terms of phosphorus.

Introduction

Seeds of many plant species are carried away by ants. Generally most seeds carried away are eaten, but a few may be dropped accidentally by the ants and thus dispersed. Even the dispersal of a few seeds in this way could produce a selection pressure opposite to that of predation: dispersal requires seed easy to find and transport, whereas escape from predation requires the reverse. The term myrmecochory, however, will be restricted to interactions where each seed bears an incentive for ants to remove it: a piece of tissue nutritionally useful to ants but unnecessary to germination or growth of the plant. Such appendages are known as food bodies, elaiosomes, or oil-bodies since they often have high lipid content. The food body is attached to the outside of the seed; its embryological origin varies and is probably not relevant to the ecology of the interaction. Typically the whole diaspore (seed plus food body) is carried towards the ant nest. If it arrives, the food body is cut or chewed off inside the nest and the seed proper is discarded either inside or outside the nest. The many observations on the natural history of myrmecochory are reviewed in English by Ridley (1930), Uphof (1948), and Van der Pijl (1972).

Most studied myrmecochores are herbs of forests or mesic meadows in the north temperate zone; *Viola odorata* provides an example. Certain adaptations are typical of these myrmecochores: the food bodies are usually soft and collapsible; the peduncles or stems supporting the infructescence often wilt or reflex as the diaspores ripen, so that the diaspores are presented in aggregations at ground level; the ant species involved commonly recruit a trail of foragers to these aggregations.

Buckley, R. C. (ed.), Ant-plant interactions in Australia.
© 1982, Dr W. Junk Publishers, The Hague. ISBN 90 6193 684 5.

Berg (1975) showed that myrmecochory is very common in Australia, and that Australian myrmecochores generally have a different suite of adaptations. About 80% of those listed by Berg (1975) are sclerophyll shrubs. Most of the remainder are perennial herbs or graminoids found in dry sclerophyll vegetation, although Davidson & Morton (1981a, b) have recently identified new myrmecochores in the arid zone. The food bodies of the sclerophyll-shrub myrmecochores are firm and durable. Their diaspores fall to the ground separately, and are often collected by ants foraging singly. The diaspores are often dispersed ballistically before being collected by ants; examples are provided by the explosive fruits of *Kennedia, Boronia* and *Petalostigma*, and by the upright pods of *Dillwynia*, from which seeds are flicked when branches whip in the wind.

Berg (1975) considered that myrmecochory was much more important in Australia than elsewhere, estimating that it possesses a total of 1 500 myrmecochore species as against 300 in the rest of the world. It is too early yet to be sure to what extent Australia is exceptional. The tropics may prove to have more myrmecochores than was thought previously (Horviz & Beattie 1980; Van der Pijl 1972). In the herbaceous understoreys of some north American deciduous forests, about 30% of species are myrmecochores (Beattie & Culver, unpublished). This percentage is comparable to that at the Australian sites described here, but the absolute number of species is greater in Australia. While it remains uncertain whether there are more myrmecochores among the sclerophyll scrubs of Australia than in any other vegetation type*, it is clear that there are far more than in the sclerophyll scrubs of California or the Mediterranean. The studies described here are directed ultimately to explaining this difference.

We have collected data of three kinds on myrmecochory in the vegetation of the West Head, N.S.W. Firstly, we classified as myrmecochores, non-myrmecochores or doubtful all the plant species (291) which appeared in 38 vegetation sample plots each 20 × 50 m. This allowed us to estimate the percentage of plant species and plant cover made up by myrmecochores, and to ask whether these percentages varied depending on site or vegetation type. These results are detailed elsewhere (Rice & Westoby 1981) and will only be summarized briefly here. Secondly, we

* See footnote to p. 86.

measured aspects of the microhabitat occupied by selected myrmecochores and non-myrmecochores in a hillside stand where the various species were interspersed. These data allow us to ask whether myrmecochores and non-myrmecochores occupy different microsites. Thirdly, we examined the costs of myrmecochory to the plant; the size of its investment in food bodies, and the currency in which this investment should be measured. The first part of this chapter presents these data, and also some anecdotal observations on the ants involved. In the second part we review published arguments explaining the high incidence of myrmecochory in Australian dry sclerophyll, assess these in the light of the available information, and present some new arguments.

Study area

The West Head is a peninsula of horizontally-bedded Hawkesbury Sandstone, about 15 × 10 km in size, in Ku-Ring-Gai Chase National Park (33° 55′ S, 151° 10′ E), just north of Sydney. The peninsula is a dissected plateau, broken slopes and steep-sided gullies alternating with flat hill tops. Soils are shallow and low in nutrients, particularly phosphorus. Annual precipitation is about 1300 mm and falls throughout the year. The vegetation cover is a mosaic of scrub, heath and eucalypt woodland formations, in the terminology of Specht (1970). Deeper moist gullies and occasional volcanic outcrops have taller forest vegetation. The vegetation is dominated by sclerophyllous plant species. Fires are frequent and most species are adapted to them in various respects.

Incidence of myrmecochores in vegetation

At well-drained sites on infertile sandstone, myrmecochores accounted for an average of 30% of the species but only 15% of the cover. Both are minimum estimates, since only undoubted myrmecochores were considered. There were far fewer myrmecochores in frequently or permanently flooded sites, or on fertile volcanic soil. With these exceptions, the incidence of myrmecochory was not related to slope or aspect of the site, or to physiognomy of the vegetation.

Individual myrmecochore species tended to be

present in more sample plots than non-myrmeco- chore species. Taken in conjunction with their ac- counting for less of the cover than of the species list, this presents a picture of the typical myrmecochore as widespread but not abundant. Over 80% of the myr- mecochore species were shrubs, although in many of these woodiness was more conspicuous in the root crown than in the stems, and the stems were often trailing. These results are documented by Rice & Westoby (1981).

Microsite preference of myrmecochores and non-myrmecochores

Methods

Microsite characteristics were examined in a species- rich (Rice & Westoby, in preparation) scrubland with emergent *Eucalyptus haemastoma* on a south- facing slope of about 15° overall. Plant names follow Beadle, Evans & Carolin (1972) throughout. Judg- ing by vegetation stature this slope had not been burnt for at least 4 years; very few seedlings of any species were present. The horizontally bedded sand- stone has eroded to a slope in which nearly flat benches up to 10 m wide alternate with sharp drops of up to 2 m. Exposed sandstone blocks have cracks of various widths and depths. Loose mineral sand and debris is distributed unevenly over this surface. The low angle of the sun with respect to the overall south slope can produce great variation in insolation between the inner and outer sides of a bench. This physical environment gives species great scope to show differentiation in their microsite preferences.

On this slope we laid out a grid of 5 × 4 squares each 10 × 10 m, located the centre of each square, and recorded several microsite characteristics for the nearest individual of each of nine selected species. No plant individual was used more than once. This ar- rangement gave us a randomly-chosen set of individ- uals of each species, without the sample for one species tending to occur higher up the slope (for instance) than the sample for another.

The species chosen were *Banksia aspleniifolia, B. ericifolia, Grevillea buxifolia, G. punicea* and *Persoonia pinifolius* (Proteaceae), and *Epacris longiflora, E. microphylla, Leucopogon esquamatus,* and *L. microphyllus* (Epacridaceae). Both families were chosen to include both myrmecochores

Grevillea spp., *Leucopogon esquamatus* and possi- bly *L. microphyllus)* and non-myrmecochores. Of the non-myrmecochores, *Banksia* spp. and *Epacris* spp. both have diaspores without fleshy parts, the seeds of *Epacris* being light, and those of *Banksia* heavy and winged.

The following variables were measured for each individual selected: (1) crown height and mean di- ameter; (2) distance to the nearest individual of the same species, measured between the two rooting points; (3) distance to the nearest tree with a diameter at breast height greater than 10 cm, assessing dbh as the sum of individual trunk diameters for multi- trunked trees; and the species, total dbh, and canopy height of this tree; (4) total percentage foliage cover of all species overshadowing its canopy; (5) percen- tage ground cover of bare rock within 0.25, 1 and 5 m radius from its rooting point; (6) depth to bedrock around the rooting point (mean of four probes); (7) litter depth around the rooting point; and (8) root habitat type. Root habitats were categorized as fol- lows: (a) rock crack less than 30 cm wide; (b) rock fissure 30 cm to 3 m wide; (c) within 6 cm of a large isolated rock, either uphill, downhill, or beside the rock; (d) pebbles or coarse gravel of 6 cm diameter or less, (e) thin sediment on the inside, middle or outside of a rock ledge; (f) shallow moss-filled hollows on ledges; and (g) deeper soil. For analysis, the height and distance of the nearest large tree were trans- formed to the angle subtended by that tree, and the distance to the nearest conspecific was expressed as number of mean crown diameters of the species con- cerned. Various sets of species were compared. The Mann-Whitney U-test was used for continuous vari- ables and the chi-square test for categorical variables.

Results

Of the different microsite variables measured, only four showed regular patterns between species; dis- tance to the nearest conspecific, percentage cover overhead (a measure of shading), mean depth to bedrock, and percentage bare rock within 1 m.

There are significant differences in the overall mic- rosite preferences of Epacridaceae and Proteaceae (Table 1); Epacridaceae occurred in deeper shade, and in shallower soils with more exposed rock nearby, than did Proteaceae. Since Epacridaceae are smaller than Proteaceae, both in general and for these species in particular, they might be expected to

Table 1. Tests of hypotheses that one subgroup of nine selected species (see text) differs in four habitat variables from another subgroup, by Mann-Whitney U-test. '0' indicates no significant difference; '+' that the first-named group is greater than the second-named at 0.05 significance level; '++' at 0.01; '+++' at 0.001; conversely with '−', '−−', '−−−'.

Subgroups	Habitat variables			
	Distance to conspecific (plant diameters)	% cover over canopy	Depth to bedrock	Bare rock within 1 m
Epacridaceae vs Proteaceae	0	+++	−−	+++
Grevillea vs *Banksia*	+++	++	0	0
Leucopogon esquamatus vs *Epacris*	+	0	0	0
Persoonia vs *Banksia ericifolia*	+++	++	0	0
Smaller-leaved vs larger-leaved epacrids	0	+	0	−−

occupy smaller root volumes and be overtopped more often. Both their stature and the microsites occupied by the two families might in turn be related to some more deep-seated differences, such as their root symbionts. In any event, they demonstrate that the methods used are in principle capable of detecting differences in microsite occupation between species.

While epacrids tended to grow closer to their nearest conspecific neighbours in absolute terms, there was no significant difference when the distance was measured in plant diameters. Plant diameters are more meaningful units here for two reasons. Firstly, they are directly related to the probability of overlap between crowns or root systems, and are in this sense a measure of intraspecific competition. Secondly, up to half a canopy diameter can be taken as a baseline distance for dispersal, being the distance a seed can achieve from the mother's rooting point if it falls vertically and is not moved subsequently.

When myrmecochores were compared with genera in the same family but lacking fleshy fruits, i.e. *Grevillea* vs *Banksia* and *Leucopogon esquamatus* vs *Epacris,* the myrmecochores were found significantly more plant diameters from their nearest conspecific (Table 1). There were no consistent differences between the physical microsites occupied by myrmecochores and by non-myrmecochores, however. The only difference found was that *Grevillea* was more shaded than *Banksia,* and this is probably attributable to *Grevillea's* lower stature.

The idea that dispersal mode affects distance to the nearest conspecific is supported by the behaviour of *Persoonia.* This genus has fleshy drupes much the size and shape of olives, which are dispersed by birds

and mammals. *Persoonia pinifolius* occurs much further from its nearest conspecific than *Banksia ericifolia* (Table 1), though both grow to a similar size, have similar narrow leaves, and are similarly killed as adults by fire, regenerating only from seed. The only habitat difference is that *Persoonia* is more heavily shaded, since it tends to grow under trees, presumably as a side-effect of bird dispersal.

Within the Epacridaceae, our selection from *Epacris* and *Leucopogon* included one species of each with very small terete leaves, less than 4 mm long × 2 mm diameter, and one of each with larger, flat leaves, more than 8 mm long × 5 mm diameter. These species were chosen on the theory that leaf size and shape can be expected to reflect differences in physiological adaptation of the adult plant to habitat. Larger- and smaller-leaved congenerics would be expected to show microsite differences associated with physiology only, without any effect of dispersal mode or other genus-level characteristics. This comparison is shown in Table 1. There were some differences in the physical habitat occupied, but not in distance to the nearest conspecific.

To summarize, the myrmecochores in this study differed from the non-myrmecochores only in being established more plant diameters away from their nearest conspecific. No consistent differences were found in their physical microsites. Although microsite differences might be found if further variables were measured, those we did measure were sufficient to show clear differences between the two families, and between different leaf-types in the Epacridaceae. The evidence suggests that in this vegetation myrmecochory increases seed dispersal distance rather than increasing seed placement into favourable microsites. It should be noticed, however, that the data do not describe seed dispersal as such, but only the fate of seeds which germinated and reached maturity.

The costs of food bodies

Methods

Diaspores from 11 species were separated into seed and food body and weighed, yielding the percent fresh weight contributed by each part. Water content is low in both seed and food body.

The above-ground parts of an individual *Acacia terminalis* were harvested at Mt. Tomah, N.S.W. in January 1979, when its fruits were ripe. It was divided into stems, leaves, pods and diaspores, and the number of ripe diaspores already shed was estimated. Dry weight, calorific and phosphorus content were measured for each component, and for the food bodies and seed separately. Calorific content was measured by micro-bomb calorimetry, and phosphorus by ashing tissues overnight at 700 °C, dissolving the ash in HCl, neutralizing with NaOH, and assaying the extract by the stannous chloride method (American Public Health Association 1965). Similar analyses were carried out for the seeds of *Patersonia sericea* (a myrmecochore), and a breakdown of dry weight and phosphorus content between the seed proper and wings or hairs was obtained for several wind-dispersed species.

Four specimens of *Acacia linifolia* at North Ryde, N.S.W., were fertilized with liquid blood-and-bone during flowering in May 1979, and four more designated as controls. Two of the controls were replaced following a bushfire which destroyed two of the fertilized plants and two controls in October 1979. All plants were harvested in November when seeds had matured and dry weight allocation to different tissues was measured.

Results

Food bodies comprise 5–10% of total diaspore weight in a range of species (Table 2). The percentage was independent of seed weight: this result disagrees with the positive correlation found over three species by Davidson & Morton (1981a).

Table 3 shows allocation of dry weight, energy and phosphorus to different aboveground biomass com-

Table 2. Mean total fresh diaspore weight, and the percentage made up by the food body, for 11 myrmecochore species. Correlation coefficient between total weight and percentage food body is 0.18, not significant.

Family Species	Total diaspore weight, mg	Food body, %
Papilionaceae		
Dillwynia floribunda	3	10
Dillwynia retorta	5	8
Bossiaea scolopendria	17	3
Iridaceae		
Patersonia sericea	5	5
Euphorbiaceae		
Micrantheum ericoides	6	4
Mimosaceae		
Acacia myrtilloides	10	7
Acacia ulicifolia	12	1
Acacia linifolia	22	10
Acacia suaveolens	35	5
Acacia oxycedrus	45	7
Acacia terminalis	48	9

Table 3. Percentage of three 'resource currencies', dry weight, energy and phosphorus, in the aboveground biomass of an *Acacia terminalis*.

	Resource currency		
	Dry weight	Energy	Phosphorus
Wood	79.6	78.2	23.2
Leaves	11.9	13.0	34.7
Pods	6.6	6.6	3.9
Diaspores	1.9	2.2	38.2

ponents in the *Acacia terminalis* individual. The striking feature of these data is that a very moderate production of seed represents an extremely large investment of phosphorus. The difference in proportional allocation is even greater than that shown in Table 3, since wood and to a lesser extent leaves represent biomass accumulated over more than one year. Hence seed production consumes much of the year's phosphorus uptake. Fertilized plants of *Acacia linifolia* increased dry weight allocation to seeds and pods about 5-fold (Table 4). Seed production is therefore more severely restricted by nutrient supply than is vegetative growth, even during the life of individual plants.

Several other lines of evidence also suggest that phosphorus is a limiting currency for seed production in many Australian sclerophyll plants. Firstly,

Table 4. Percentage of dry weight in different tissues of fertilized and unfertilized *Acacia linifolia*, ± one s.d., and probability that allocation is different between fertilized and unfertilized plants (*t*-test).

Tissue	Unfertilized ($n = 4$)	Fertilized ($n = 2$)	p
Wood	71 ± 10	64 ± 3	0.40
Roots	10 ± 2	13 ± 6	0.32
Leaves	17 ± 9	11 ± 2	0.42
Pods	1.6 ± 0.9	8.8 ± 1.1	0.0009
Seeds	0.9 ± 0.5	4.3 ± 0.07	0.0009
Pods + seed	2.5 ± 1.4	13 ± 1.1	0.0008

Table 5. Percentage of different 'resource currencies', dry weight, energy and phosphorus, found in different parts of diaspores of two myrmecochores and five wind-dispersed species.

	Resource currency		
	Dry weight	Energy	Phosphorus
Acacia terminalis			
seed	91.1	91.1	98.7
food body	8.9	8.9	1.3
Patersonia sericea			
seed	89.9	89.9	97.0
food body	10.1	10.,1	3.0
Hakia sericea			
seed	95.7		99.4
wing	4.3		0.6
Banksia aspleniifolia			
seed	91.8		99.6
wing	8.2		0.4
Banksia ericifolia			
seed	90.5		98.5
wing	9.5		1.5
Banksia serrata			
seed	98.5		98.6
wing	11.5		1.4
Petrophile fucifolia			
seed	81.5		98.8
hairs	18.5		1.2

soil phosphorus levels are very low, with typical values of 20–40 ppm total phosphorus for soils on the Hawkesbury Sandstone (Hannon 1956). Secondly, sclerophylly is generally associated with low-phosphate soils in Australia (Beadle 1953, 1954, 1962, 1966, 1968) although elsewhere it is a response to summer drought in Mediterranean-type climates (Di Castri & Mooney 1973). Thirdly, phosphorus fertilization of sclerophyll vegetation has often been reported to 'accelerate the life cycle' and produce earlier reproduction in sclerophyll species, unless it damages them by phosphorus toxicity (Connor & Wilson 1968; Heddle & Specht 1975; Jones 1968; Specht 1963, 1975; Specht, Connor & Clifford 1977). Finally, seedlings of many species can grow to many times the dry weight of their seeds – 30 times for *Banksia grandis* Willd., for example – using phosphorus from seed reserves only, without any uptake from the soil (Barrow 1977). This independence of the seedling from root uptake during early growth, perhaps before root symbionts are established, is probably the advantage obtained by the great concentration of phosphorus in the seed.

Overall, therefore, it seems probable that phosphorus is the limiting currency for seed production in these sclerophyll plants. This is relevant to the evolution of dispersal since the food bodies examined contain relatively little phosphorus (Table 5). On the other hand, the wings or hairs of species adapted for wind dispersal also cost little phosphorus. If energy or dry weight were the limiting currency, the species shown could increase seed production about 10% by dispensing with adaptations for dispersal and reallocating their resources. If phosphorus were the limiting currency, however, reallocating the resources allocated to dispersal would increase seed production by less than 3%.

The ants involved

Ants participate in the interaction because the benefits of eating the food bodies outweigh the costs of finding and carrying the diaspore. From the ants' point of view the interaction is a problem in optimal foraging. We doubt that the food bodies are such a major resource for the ants that any special adaptations are selected for: our non-quantitative observations indicate that most of the ants involved are general foragers, taking a broad spectrum of food types. On the West Head most seeds were taken by *Rhytidoponera victoriae* or by *Pheidole* sp. 'B' of the CSIRO Division of Entomology. Occasionally seeds were taken by an *Iridomyrmex* species. Each species took a wide range of seed sizes. On the West Head at least, myrmecochory does not involve close coevolution between particular species of plants and ants.

Conclusions

The data described above, together with published information and our qualitative observations, present quite a coherent picture of myrmecochory as a phenomenon in the vegetation of the West Head. It is a major means of dispersal, ranking with wind dispersal, for instance, in its quantitative importance. It is uncommon only on fertile soils, or where flooding limits the ant fauna.* Tight coevolution between particular plant and ant species is not involved. The benefit to the plants appears to be simple distance of dispersal, rather than placement. Compared to non-myrmecochores, myrmecochores tend to occur widely through the vegetation, but as scattered individuals at relatively low density. We believe this sort of spatial distribution is part of a life-style which typically also involves good dispersal generally, rather than myrmecochory specifically.

Discussion - costs and benefits of myrmecochory to the plant

Various arguments have been advanced regarding the benefits of myrmecochory. The following sections assess these arguments in a systematic order. Firstly, we indicate general problems in explaining why particular dispersal methods should be used under particular ecological circumstances, and define more precisely what is to be explained in the case of myrmecochory. Secondly, we assess the possible benefits and costs of myrmecochory: any characteristic which is subject to natural selection can be favoured by decreased cost as easily as by increased benefit. Thirdly, we assess ecological circumstances which might change costs and benefits relative to those of alternative methods of dispersal; and finally, we summarize the most probable controls for the observed distribution of myrmecochory. Throughout this discussion, phrases such as 'the plant chooses to . . .' are used as shorthand for 'those plants which did . . . survived and reproduced better than those which did not . . . with the result that over evolutionary time that the plants now alive do . . .'.

On dispersal biology

Why should a particular vegetation use a particular mixture of dispersal strategies? The 'why' in this

* M.C. Ricketts (unpublished data, July 1980) has found only one *Camponotus* sp. in wet heaths, compared to 12 ant species in quadrats in adjacent dry sclerophyll vegetation.

question can be referred to 'ecological' or 'evolutionary' time. The first is the sense used here: 'what ecological circumstances change the costs or benefits so as to favour a given phenomenon more in one place than in another?' The second possible meaning is 'what ecological circumstances favoured the early evolution of the phenomenon more in one place than in another?' Shine & Bull (1979) have emphasized that situations which favour a character's retention do not necessarily favour its emergence, or vice versa. Where this is true, speculative reasoning is possible, but tests are difficult, because the critical test situations are usually in the past. Fortunately it seems unlikely that there was any major adaptive valley to be crossed in the emergence of myrmecochory. Food bodies are enlargements of structures such as the strophiole which are already present, and it seems likely that even small food bodies would somewhat increase the probability that an ant would carry the seed.

A plant's choice of a dispersal method is closely tied to two other choices that it makes through evolutionary time. First, it must choose what size of seed it will produce. Although dispersal can be a factor in this decision, seed size is selected very powerfully by factors affecting seed and seedling survival. Seed size in turn affects what dispersal methods can be used. Small seeds are predisposed to wind dispersal by their already large surface area–weight ratio. On the other hand, the investment needed to persuade an animal vector to go out of its way to collect a seed is a smaller percentage increase in the cost of a large seed than in that of a small one.

Second, the plant must choose how much distance it wishes its diaspores to achieve, or to be precise, what frequency distribution of dispersal distances it wishes to achieve. For a given seed weight, a longer dispersal distance will generally call for a greater investment in special structures on the diaspore. The relationship of fitness to distance achieved is therefore important. If fitness rises for (say) the first 2 m, but does not change thereafter, the large investment in a fleshy food body needed to use inherently long-range dispersal agents such as birds would be pointless.

Thus the mixture of dispersal methods found in vegetation can vary for three types of reason: first, because the mixture of seed sizes is different, and this favours different dispersal methods: second, because different advantages are being achieved by dispersal, e.g. a premium on dispersal beyond 10 m in one case and not in another; or third, because even though seeds of the same size are being dispersed to the same distances, the relative economy with which two methods achieve this varies between two situations.

Ideally, the dispersal biology of different vegetation types should be characterized in terms of their spectrum of seed weights, distances achieved, and dispersal methods used. With this information, it would be possible to tell whether a difference in the incidence of myrmecochory represented a difference in the cost-effectiveness of myrmecochory compared to some other dispersal method (and if so which) for achieving a particular dispersal result, or rather was a side-effect of differences in the seed sizes or dispersal distances between the vegetation types. Until such extensive information on dispersal accumulates, we must interpret those differences we know of as best we can.

What is to be explained?

Discussions of the benefits of myrmecochory have either asked in what respect it is adaptive in general, or what favours it in Australia compared to other continents. However, it seems clear that the two syndromes of myrmecochory, in herbs of deciduous forests and in sclerophyll shrubs, are selected in different contexts of seed sizes and benefits to be obtained by dispersal over various distances. The question we will consider is: why is myrmecochory found more in sclerophyll shrubs than in other growth-forms in Australia, but not in North America or the Mediterranean? In the following discussion of possible benefits and costs, it will become apparent that many proposed explanations can account for a concentration of myrmecochory in Australia, but not for its distribution in sclerophyll shrubs there; or for a concentration of myrmecochory in sclerophyll shrubs, but not for this occurring in Australia but not elsewhere.

Benefits

It should be borne in mind that a benefit is only a probability of increased fitness. This probability will not be manifested for each seed. Indeed, given that for most plant species somewhere between 1 in 10^2 and 1 in 10^6 of the seeds produced will establish and grow to maturity (Williams 1975; Harper 1977), the

fate of the great majority of seeds may be irrelevant to the merits of an adaptation. On the other hand most of the costs of myrmecochory must be incurred for every seed which is offered to the dispersal agent.

The first major benefit of dispersal may be increased distance between progeny and parent, reducing competition from the parent. It does not necessarily reduce competion between seedlings, since ant dispersal can produce very heavy concentrations of seed in or around ant nests. If the area outside the mother's canopy is occupied by other adults, dispersal will not reduce the competition suffered by the seedling, but it will reduce competition between kin: success becomes variable-sum for an allele, even though it is fixed-sum for the species population. This could also apply to seed dispersal into high density patches around ant nests. Distance may also reduce predation, since predation can be greater under adults or in high-density patches (O'Dowd & Hay 1980; Reichman & Oberstein 1977). Predation could here include species-specific or genotype-specific pathogens, which may accumulate in the soil under adult plants (Fox 1977).

Different dispersal methods could produce the same mean dispersal distances but different frequency distributions. Wind dispersal might produce a more or less Gaussian distribution with a mean of 3 m; ballistic dispersal might move all seeds at least 2 m but none more than 5 m; fruit-eating birds might regurgitate most of the seeds immediately beneath the mother, but disperse a minority to 20 m or more. If seeds established far from the mother have very high reproductive success, selection can favour a mechanism which disperses a few seeds to long range even if most achieve no dispersal at all. If fitness is increased mainly by removal beyond the mother's canopy, however, the mother (Westoby 1981) will be selected to use dispersal such as a ballistic mechanism, or wind dispersal from an elevated infructescence, which achieves this moderate range for most seeds. The overall relation of fitness to distance may include an abrupt increase in fitness at the margin of the mother's influence; a gradual increase proportional to the decreased density of other seeds from the same mother; or an increase proportional to the probability that any recently-opened patches will be reached by one or more seeds. The last two effects will be roughly proportional to the square of the distance achieved. Qualitative observations on the seed shadows produced by myrmecochory suggest that nearly all myrmecochore seeds are found and carried some distance by ants, but few travel more than 5 m. Probably myrmecochory should be favoured where fitness increases with distance achieved in such a way as to reward reliable but moderate dispersal.

The second major benefit of dispersal may be to place seeds selectively with respect to small-scale environmental pattern. It has been variously suggested or shown that myrmecochory may place seeds at a depth in the soil which protects them from fire, predation, or midsummer heat (Berg 1975; Shea, McCormick & Portlock 1979), on rotting logs (Handel 1976), or in the nutrient-rich soil of ant nests (Culver & Beattie 1978; Berg 1975; Davidson & Morton 1981a, b). We have observed two further sorts of placement. Firstly, a proportion of myrmecochore seeds collected by ants never reach ant nests. O'Dowd & Hay (1980) found that 10% of *Datura discolor* seeds removed by the harvester ant *Pogonomyrmex badius* were in this category. Because many myrmecochore seeds have a hard smooth testa, ants can not carry them without the food body to use as a handle. Further, the food body is often not very firmly attached. It comes off as a whole after it has been used to pull the seed over an obstruction a few times, and then ants do not carry the seed any further. On the West Head such seeds often finish up wedged against a stone or under a twig; this may help root penetration during seedling establishment. It may be adaptive that many food bodies come off as wholes rather than partially when pulled, and that they are attached to the seed with intermediate firmness. Seeds also fail to reach ant nests when a small ant species recruits a foraging group to a relatively large seed (e.g. *Pheidole* sp.'B' to *Acacia terminalis).* Each ant then bites a small piece off the food body, which is gradually nibbled down to nothing, at which point the seed is abandoned. The seed is rarely moved more than a few cm; this sort of event is apparently a failure of the dispersal strategy. Secondly, seeds that do reach the nest are often discarded on bare, compact soil near a nest exit; during storms the seeds are then washed into accumulations of organic debris which may be favourable sites for seedling establishment.

A third benefit proposed is that ants biting the surface of seeds they disperse have been shown to increase germinability (Culver & Beattie 1978). This could enhance the benefit of dispersal if there is an advantage in seeds germinating only after ants have moved them.

It is relatively rare for a plant species to use more than one agent to disperse any given seed (Van der Pijl 1972), since the morphological adaptations required are generally incompatible. A fourth possible benefit of the small compact food body of myrmecochores is that they are compatible with ballistic dispersal; this combination is common in Australian myrmecochores. It is presumably an improvement on either mechanism alone, though there are no quantitative data on the advantages of investment in a second dispersal system rather than increased investment in the first, and the two could conflict if ants carried ejected diaspores back towards their source. In addition, ballistic dispersal would not be compatible with aggregated seed presentation to trail-forming ants, common in northern hemisphere myrmecochores. These factors are considered in detail elsewhere (Westoby & Rice 1981) but seem unlikely to explain the concentration of myrmecochory in Australia.

Costs of myrmecochory

The costs of myrmecochory include the costs of the food body itself, the cost of rendering the seed inedible and the costs of seeds lost by poor placement. Many food bodies have high lipid and hence energy content, although this is variable (Bresinsky 1963, cited by Van der Pijl 1972; Marshall et al. 1979; O'Dowd & Hay 1980). Evidence given above shows that the food bodies of at least two sclerophyll species cost much less in terms of phosphorus than of energy. In general, a plant will be selected to concentrate in food bodies resources valuable to the ants, but to invest as little as possible of resources limiting to the plant's other activities. Myrmecochory will be favoured where these two objectives can be achieved simultaneously – that is, where different resources are limiting for the plants and the ants.

Myrmecochory is not a viable strategy unless the seed itself is protected from being eaten by the ants. It has therefore been suggested that myrmecochory will not occur in ecosystems where harvester ants are common (Van der Pijl 1972; Harper 1977). It has been shown recently that the myrmecochores *Dendromecon rigida* and *Datura discolor* are routinely dispersed by the harvester ant *Pogonomyrmex badius* (Bullock 1974; O'Dowd & Hay 1980), but these appear to be exceptional. The seeds of *Datura discolor* are protected by a strong testa, and are carried to the nest but then expelled intact. The cost of this testa is acceptable since the diaspores themselves are exceptionally large and energy-rich (Reichman 1975). A hard testa is also the most common form of seed protection in Australian plants, being found in most Mimosaceae, Papilionaceae, Rutaceae, Dilleniaceae and Euphorbiaceae. By physical prevention of water uptake it imposes dormancy which is commonly broken by fire. It is thus primarily a fire adaptation, and its costs are necessarily incurred independent of myrmecochory in many Australian sclerophyll taxa.

The costs of poor seed placement by myrmecochory have generally been ignored. Ants may place seeds in favourable microsites, but it is perhaps equally likely that they will abandon seeds in sites less favourable than those where they fell to ground initially. Seeds may be deposited on flat hard surfaces outside nests or stored in ant nests too deep to emerge on germination. They may also be aggregated so that seedling competition is increased (Shea et al. 1979). To assess whether placement by ants is a cost or a benefit, explicit measurements of the fitness of seeds in different microsites are needed; none are yet available.

Possible explanations

To restate the problem: why is myrmecochory common in sclerophyll shrubs compared to other growth-forms in Australia, but not in other continents?

Variations in myrmecochory from place to place or from growth-form to growth-form can be accounted for only in terms of costs or benefits which vary similarly. Possible arguments can be divided into three groups, as follows.

Firstly, it could be argued that the two main probable benefits of myrmecochory, distance and placement, are for some reason more useful to plants in Australian dry sclerophyll vegetation than elsewhere; or if placement is a disadvantage, that it is less so in Australian dry sclerophyll. The benefits of seed dispersal distance certainly do vary between different vegetation types and plant growth-forms. Though there are no quantitative data, it seems likely that seeds in sclerophyll shrublands will gain a substantial advantage from dispersal beyond the direct influence of the mother, but no further advantage from greater distances. Such a benefit–distance function would favour the evolution of dispersal systems which transport most seeds a few metres, and myrmecochory is one such system. This argument could explain a high incidence of myrmecochory in sclerophyll shrubs, but not why such a pattern should be found in Australia but not in North America or the Mediterranean.

Similar problems apply to arguments involving placement. It has been argued that burial provides protection from fire (Berg 1975) or midsummer heat (Shea et al. 1979). But summer midday soil temperatures are no higher in Australia than elsewhere at comparable latitudes, or in sclerophyll vegetation than in the arid zone; and Californian and Mediterranean shrublands have regimes of frequent fire.

Secondly, it could be argued that a diaspore with given cost to the plant is more likely to be beneficially dispersed in Australian dry sclerophyll than elsewhere because of differences in the ant fauna (Berg 1975). It is well known that the Australian ant fauna is unusually rich (Taylor 1972).

Two questions need to be asked about this explanation. First, is the richness of the Australian ant fauna concentrated, like myrmecochory, in dry sclerophyll vegetation? In most parts of the world ant faunas are richest in humid vegetation, but in Australia the richness is found particularly in semi-arid and dry sclerophyll vegetation (Taylor 1972). In this respect the richness of the ant fauna corresponds to the distribution of myrmecochory, provided some other explanation can be found for the relative shortage of myrmecochory in Australian semi-arid vegetation. The presence of harvester ants in the semi-arid vegetation could supply this other explanation.

The second question is: Why should a large number of ant species encourage the development of myrmecochory? On the West Head, myrmecochory does not involve coevolution between particular ant species and particular plant species. Most diaspores of all the plant species we have observed are removed by only two common ant species, each of which takes a large range of diaspore sizes. There seems no reason why the availability of many kinds of ant should encourage myrmecochory. It may be that the richness of ant species reflects high absolute abundance of ants; that is, the prospect that a diaspore will be found by an ant is great. We know of no applicable data on the absolute abundance of ants. But even if ants are especially abundant in Australian dry sclerophyll vegetation, it is not entirely clear that this is an important difference from the point of view of the plant. Studies in the north temperate zone usually find that a majority of diaspores are taken by ants within 24 hours of maturity. (This is reflected in the ephemeral nature of many of their food bodies.) Even in the north temperate, then, the probability that a diaspore which remains desirable to ants will go unfound for an extended period is vanishingly small (e.g. for 100 days less than 0.5^{100} or 8×10^{-31}). The risk is not one of going undetected by ants, but of being found by a predator (rodent or bird) before being moved by an ant (O'Dowd & Hay 1980; A. J. Beattie, personal communication). This risk is probably lower in Australian dry sclerophyll vegetation. Seed-eating rodents are markedly less abundant there than in Europe or North America (Morton 1979); also the long-lasting form of food body adopted by Australian myrmecochores suggests that they have not been under pressure to be found immediately by ants. For these reasons the idea that myrmecochores are common in Australian dry sclerophyll vegetation because of the species-rich ant fauna there seems to us to have serious difficulties.

The third category of possible explanations is that myrmecochory is less expensive, for a given benefit, in Australian dry sclerophyll vegetation than elsewhere. We have suggested two senses in which myrmecochory may be less expensive in Australian dry sclerophyll. Firstly, this highly fire-adapted vegetation contains many species which would in any event have invested in heavy testas. They therefore incur no extra cost for protecting the seed from ant consumption. This could explain a concentration in sclerophyll vegetation, since most sclerophyll floras are highly adapted to fire, but it is inconsistent with the shortage of myrmecochores among sclerophyll shrubs in California and the Mediterranean. Secondly, the distribution of Australian sclerophyll shrublands, unlike those on most other continents, is bounded by the distribution of infertile soils. Phosphorus may be the essential currency in which investment in seed production and dispersal should be measured, and myrmecochory is cheap in this currency. At least some food bodies of myrmecochores in this vegetation contain relatively little phosphorus, but are effective as incentives for the ants. This reduction in the effective cost of food bodies could explain both the concentration of myrmecochory in sclerophyll shrubs within Australia, and its relative absence

from sclerophyll shrubs in North America and California.

It is not only food bodies intended for ants, however, which become relatively cheap when phosphorus is the currency. Food bodies for birds or rodents and rigid structures to increase air resistance or impel seeds ballistically also can be constructed using little phosphorus. When phosphorus is the currency, it is therefore dispersal as a whole that is cheap compared to lack of dispersal, rather than myrmecochory that is cheap compared to other methods of dispersal. Other arguments, such as those about the benefits of achieving moderate distances for all diaspores in sclerophyll shrublands, need to be invoked to explain why increased dispersal should take the form of myrmecochory and ballistic devices. One consequence of this hypothesis is that if Australian sclerophyll shrubs were compared with (say) Californian sclerophyll shrubs of similar seed weight, a lower incidence of myrmecochory in California should not be compensated for by a higher incidence of some other means of dispersal.

Likely explanations

We can summarize the above account of factors which might affect the costs or benefits of myrmecochory as follows. Many of the proposed benefits of myrmecochory seem very unlikely to be greater either in Australia than on other continents, or among sclerophyll shrubs than other growth-forms. Fire regimes, producing pre-adaptation for heavy testas and selection for seed burial, could explain the concentration of myrmecochory in sclerophyll shrubs, but not the differences between continents. The distribution of ant species richness fits that of myrmecochory both between continents and within Australia, provided harvester ants explain the relative absence of myrmecochory from the Australian semi-arid, but it is far from clear why a species-rich ant fauna should encourage myrmecochory. Finally, the cheapness of myrmecochory, among other dispersal methods, where phosphorus is the limiting resource for seed production might explain both its commonness in sclerophyll vegetation in Australia and its relative absence from sclerophyll vegetation in California and the Mediterranean, provided subsidiary reasons could be found to explain why myrmecochory rather than other dispersal methods should be favoured by low-phosphorus soils. This explanation is also consistent with our data suggesting that myrmecochory generally increases spacing within a species, without being of especial use to species using particular topographic sites or particular microsites. Given the limited and circumstantial evidence available, we feel that the most likely explanation for the concentration of myrmecochory in Australian sclerophyll vegetation is the low phosphorus soils which characterize that vegetation. Under this hypothesis a concentration of myrmecochory would also be expected in the Cape Flora of South Africa. R. Y. Berg (personal communication) and the descriptions in Dyer (1975), suggest that food bodies of the type associated with myrmecochory in Australia may indeed be quite common in South Africa. This possibility deserves investigation.*

References

American Public Health Association, 1965. Standard Methods for the Examination of Water and Waste Water, 12th edn.

Barrow, N. J., 1977. Phosphorus uptake and utilization by tree seedlings. Aust. J. Bot. 25: 571–584.

Beadle, N. C. W., 1953. The edaphic factor in plant ecology with a special note on soil phosphate. Ecology 34: 426–428.

Beadle, N. C. W., 1954. Soil phosphate and the delimitation of plant communities in eastern Australia. Ecology 35: 370–375.

Beadle, N. C. W., 1962. Soil phosphate and plant communities in eastern Australia II. Ecology 43: 81–288.

Beadle, N. C. W., 1966. Soil phosphate and its role in molding segments of the Australian flora and vegetation with special reference to xeromorphy and sclerophylly. Ecology 47: 991–1007.

Beadle, N. C. W., 1968. Some aspects of the ecology and physiology of Australian xeromorphic plants. Aust. J. Sci. 30: 348–355.

Beattie, A. J. & Culver, D. C., 1981. The guild of myrmecochores in the herbaceous flora of West Virginia forests. Ecology 62: 107–115.

Berg, R. Y., 1975. Myrmecochorous plants in Australia and their dispersal by ants. Aust. J. Bot. 23: 475–508.

Bresinsky, A., 1963. Bau, Entwicklungsgeschichte und Inhaltstoffe der Elaiosomen. Biblioth. Botan. 126: 1–54.

Bullock, S. H., 1974. Seed dispersal of Dendromecon by the seed predator Pogonomyrmex. Madrono 22: 378–379.

Connor, D. J. & Wilson, G. L., 1968. Response of a coastal Queensland heath community to fertilizer application. Aust. J. Bot. 16: 117–123.

Culver, D. C. & Beattie, A. J., 1978. Myrmecochory in Viola: dynamics of seed–ant interactions in some West Virginia species. J. Ecol. 66: 53–72.

Davidson, D. W. & Morton, S. R., 1981a. Myrmecochory in some plants (F. Chenopodiaceae) of the Australian arid zone. Oecologia 50: 357–366.

Davidson, D. W. & Morton, S. R., 1981b. Competition for dispersal in ant-dispersed plants. Science 213: 1259–1261.

Di Castri, F. & Mooney, H. A. (eds.), 1973. Mediterranean-Type Ecosystems. Springer, Berlin, 500 pp.

Docters van Leeuwen, W. M., 1929. Kurze Mitteilung über Ameisen-Epiphyten aus Java. Ber. dtsch. Bot. Ges. 47: 90–99. (Cited by Van der Pijl, 1972).

Dyer, R. A., 1975. The Genera of Southern African Flowering Plants. Vol. 1. Dicotyledons. Republic of South Africa Dept. Agric. Tech. Services, 756 pp.

Fox, J. F., 1977. Alternation and coexistence of tree species. Am. Nat. 111: 69–89.

Handel, S. N., 1976. Dispersal ecology of Carex pedunculata (Cyperaceae), a new north American myrmecochore. Am. J. Bot. 63: 1071–1079.

Hannon, N. J., 1956. The status of nitrogen in the Hawkesbury sandstone soils and their plant communities in the Sydney district. I. The significance and level of nitrogen. Proc. Linn. Soc. N.S.W. 81: 119–143.

Harper, J. L., 1977. Population Biology of Plants. Academic Press, London.

Heddle, E. M. & Specht, R. L., 1975. Dark Island Heath, (Ninety-Mile Plain, South Australia). VIII. The effect of fertilizers on composition and growth, 1950–1972. Aust. J. Bot. 23: 151–164.

Horvitz, C. C. & Beattie, A. J., 1980. Ant dispersal of Calathea (Marantaceae) seeds by carnivorous ponerines (Formicidae) in a tropical rain forest. Am. J. Bot. 67: 321–326.

Jones, R., 1968. Productivity studies on heath vegetation in southern Australia. The use of fertilizers in studies of production processes. Folia Geobot. Phytotax. (Praha) 3: 355–362.

Marshall, D. L., Beattie, A. J. & Bollenbacher, W. E., 1979. Evidence for diglycerides as attractants in an ant–seed interaction. J. Chem. Ecol. 5: 335–344.

Morton, S. R., 1979. Diversity of desert-dwelling mammals: a comparison of Australia and North America. J. Mammal. 60: 253–264.

O'Dowd, D. J. & Hay, M. E., 1980. Mutualism between harvester ants and a desert ephemeral: seed escape from rodents. Ecology 61: 531–540.

Reichman, O. J., 1975. Relationships between dimensions, weights, volumes, and calories of some Sonoran Desert seeds. Southwest Naturalist 20: 573–575.

Reichman, O. J. & Oberstein, O., 1977. Selection of seed distribution types by Dipodomys merriami and Perognathus amplus. Ecology 58: 636–643.

Rice, B. L. & Westoby, M., 1981. Myrmecochory in sclerophyll vegetation of the West Head, N.S.W. Aust. J. Ecol. 6: 291–298.

Rice, B. L. & Westoby, M., (in prep.). Species richness and composition of sclerophyll vegetation of the West Head, N.S.W.

Ridley, H. N., 1930. The Dispersal of Plants Throughout the World. Reeve, Ashford, Kent.

* Since we wrote this paper (in April 1980) N.W. Bond (personal communication) and his colleagues have identified about 1 000 myrmecochores in South Africa, almost entirely from the fynbos on low-nutrient soils (Milewski & Bond, this volume). This represents a concentration of myrmecochory comparable to that in Australia.

Shea, S. R., McCormick, J. & Portlock, C. C., 1979. The effect of fires on regeneration of leguminous species in the northern jarrah *(Eucalyptus marginata* Sm) forest of Western Australia. Aust. J. Ecol. 4: 195–205.

Shine, R. & Bull, J. J., 1979. The evolution of live-bearing in lizards and snakes. Am. Nat. 113: 905–923.

Specht, R. L., 1963. Dark Island Heath (Ninety-Mile Plain, South Australia). VII. The effect of fertilizers on composition and growth, 1950–1960. Aust. J. Bot. 11: 67–94.

Specht, R. L., 1970. Vegetation. In: Leeper, G. W. (ed.), The Australian Environment. 4the edn. Melbourne University Press, Melbourne.

Specht, R. L., 1975. The effect of fertilizer on sclerophyll (heath) vegetation. Search 6: 459–461.

Specht, R. L., Connor, D. J. & Clifford, H. T., 1977. The heath-savannah problem: the effect of fertilizer on sand-heath vege-tation of North Stradbroke Island, Queensland. Aust. J. Ecol. 2: 179–186.

Taylor, R. W., 1972. Biogeography of insects of New Guinea and Cape York Peninsula. In: Walker, D. (ed.), Bridge and Barrier: The Natural and Cultural History of Torres Strait, 213–230. Aust. Nat. Univ., Res. Sch. Pacific Studies, Pub. BG/3.

Uphof, J. C. Th., 1942. Ecological relations of plants with ants and termites. Bot. Rev. 8: 563–598.

Van der Pijl, L., 1972. Principles of Dispersal in Higher Plants. 2nd edn. Springer-Verlag, Berlin.

Westoby, M., 1981. How diversified seed germination behavior is selected. Am. Nat. 118: 882–885.

Westoby, M. & Rice, B., 1981. A note on combining two methods of dispersal-for-distance. Aust. J. Ecol. 6: 189–192.

Williams, G. C., 1975. Sex and Evolution. Princeton University Press, Princeton.

CHAPTER NINE

Convergence of myrmecochory in mediterranean Australia and South Africa

A. V. Milewski and W. J. Bond

Abstract. Interactions between ants and plants were compared on sandy soils under a mediterranean climate in Australia and southern Africa. Both the Barrens in Western Australia and the Caledon coast in the southwestern Cape, South Africa, support fire-prone sclerophyllous shrublands. The plant community on siliceous, leached, acidic soils in South Africa was found to have a high incidence of dispersal of diaspores by ants, despite differences in the fauna and flora between the continents. This indicates evolutionary convergence in response to similar physical environments. Comparison of different substrates under the same climates suggests that the poverty of potassium and other soil nutrients favours myrmecochory over other strategies, such as dispersal of seeds by birds eating large fleshy fruits, which require relatively fertile soils and protection from fire. Further studies of the relationship between soil nutrient status and patterns of plant dispersal are needed.

Introduction

The hypothesis implicit in studies of evolutionary convergence is that the ecological characteristics of plants and animals and the communities they form are ultimately determined by the physical environment. This hypothesis can be tested by selecting two or more regions with long periods of evolution in isolation from each other and with as similar physical environments as possible (Mooney 1977; Orians & Solbrig 1977). Convergence in the morphology, function or community organization of different biotas is evidence for the role of physical environment in evolution (Schimper 1903; Mooney & Dunn 1970; Parsons & Moldenke 1975; Parsons 1976). Non-convergence in the biotas of matched study areas may be due to the differences in the history of the areas or to phylogenetic constraints on evolutionary plasticity (Cody & Mooney 1978; Pianka 1979; Walter 1979).

Adaptations for seed dispersal are a complex compromise involving several stages in the life history of a plant. Integration between a number of plant organs is required for successful seed development, dispersal and seedling establishment (Van der Pijl 1969; Stebbins 1971). For example, regeneration of trees in temperate woodlands depends on seed size, which implies constraints on seed number and possible dispersal mechanisms, and influences predation of seed before and after dispersal (Salisbury 1942; Janzen 1969; Harper et al. 1970). Perhaps because of this complexity, few attempts have been made to relate suites of dispersal strategies to physical environment (Danserau & Lems 1957) and none, as far as we are aware, to assess convergence of dispersal strategies on different continents.

Australia has an extraordinarily high incidence of plants with seed dispersed by ants. Most of these myrmecochorous species are endemic, suggesting evolution of the syndrome in situ after isolation of the flora (Berg 1975). Most of the plants dispersed by ants occur in fire-prone shrublands or woodlands with shrubby understoreys on infertile soils, and they are rare in rainforests on richer soils (Berg 1981; Clifford & Drake 1981; Rice & Westoby 1981). Similar fire-prone shrublands, termed Cape fynbos, occur on acid, sandy soils in the southwestern and southern Cape Province of South Africa. This study compares

Buckley, R. C. (ed.), Ant-plant interactions in Australia.
© 1982, Dr W. Junk Publishers, The Hague. ISBN 90 6193 684 5.

ant and plant communities in small areas of Western Australia and South Africa with particularly similar mediterranean climates. Two contrasting sites, one with nutrient-poor and one with relatively nutrient-rich soils, were examined on each continent. We test the degree of convergence in dispersal by ants and other agents such as birds, and interpret dispersal in terms of reigning environmental factors, particularly soil fertility. We regard dispersal strategies as convergent where they differ less in sites of similar fertility between continents than in sites of different fertility within a continent.

Selection of matched sites: the Barrens and the Caledon coast

Australia and South Africa are particularly suited to testing the relative roles of biotic and physical controls on evolution. Similarities in the physical environment (Milewski 1979) allow very close matching of study sites, yet the continents have been separated since the break-up of Gondwanaland in the late Cretaceous, at a relatively early stage in the evolution of angiosperms (Raven & Axelrod 1974). The Capensis flora of the extreme southwestern tip of the African continent is distinct enough to be recognized as one of the world's six floral kingdoms (Good 1974) although it occupies less than 4% of South Africa and

only a tiny fraction of the world's land area (Goldblatt 1978; Taylor 1978). Southwestern Australia is similarly rich in species and endemic taxa (Hopper 1979) and forms a distinct floristic province (Beard 1981). Despite long periods of isolation, however, the two regions share a number of southern taxa (e.g. Proteaceae, Restionaceae), so that the question of common phylogenies cannot be entirely excluded when assessing convergence.

Particularly similar mediterranean climates and soils can be found at the Barrens, between Albany and Esperance in Western Australia (34 °S, 120 °E), and the Caledon coast in the southwestern Cape of South Africa (34° 30′S, 19° 30′E). The Barrens is a windy, hilly part of mediterranean Australia with relatively frequent, light precipitation (Fig. 1). The Caledon coast is typical of the Cape fynbos area. Abrupt low mountains of acidic sedimentary rocks, flanked by sand plains and dunes, are found in both regions (Spies et al. 1963; Walsh 1968; Aplin & Newbey 1982).

On each continent we chose one study site, the infertile site, on leached, siliceous sand at the base of a quartzite or sandstone hill near the coast. These sites were very similar in topography, soils and apparently also microclimate (Plate 1). Soil nutrient concentrations are listed in Table 1. The Caledon coast soils were slightly more fertile than those in

Table 1. Nutrient status of soils in the study sites at the Barrens, Western Australia, and the Caledon coast, South Africa. All the soils were sands, well-drained throughout the year to a depth of at least 1 m. Values are in parts per million. Exchangeable cations were determined using $1 N$ NH_4Cl. Nitrogen values for the South African sites are estimates from similar vegetation farther afield.

	Australia		South Africa	
	Infertile site	Fertile site	Infertile site	Fertile site
Total phosphorus				
surface	18	60	95	240
subsoil (60 cm)	10	30	75	180
Exchangeable potassium				
surface	12	20	16	95
subsoil	20	12	4	38
Exchangeable calcium				
surface	240	2600	240	6750
subsoil	128	2000	80	4040
Exchangeable magnesium				
surface	29	240	28	370
subsoil	20	120	5	120
Exchangeable sodium				
surface	5	34	18	150
subsoil	5	22	19	120
Total nitrogen				
surface	200	1400	600	1800
subsoil	100	500	300	700

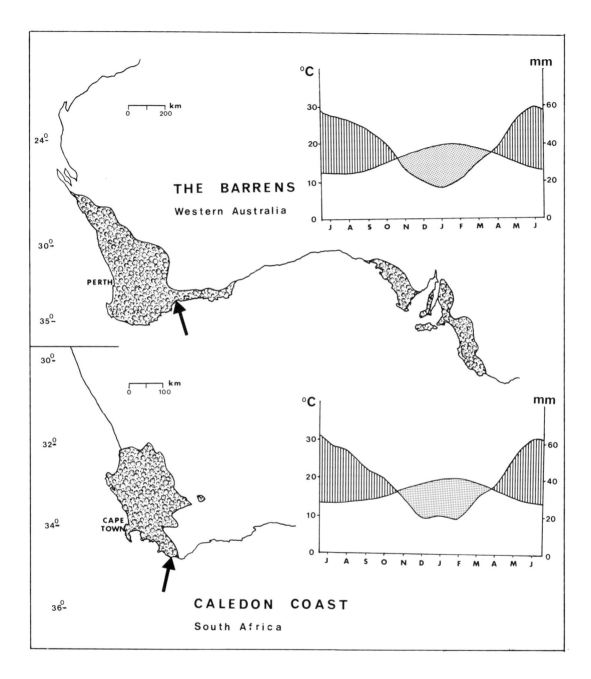

Fig. 1. Climate at the Barrens and the Caledon coast: ombrothermic diagrams for the most similar climates in mediterranean Australia and southern Africa, following Milewski (1979). Mean monthly temperatures are given on left scale and mean monthly precipitation on right scale. The diagrams represent the study sites, receiving mean annual precipitation of 470 mm on both continents. Maps show location of the Barrens and the Caledon coast (arrowed) and the area of mediterranean climate on each continent (shaded) after Milewski (1979).

Plate 1. General view of the study site on infertile soil at the Caledon coast, southwestern Cape, South Africa, showing the Cape fynbos vegetation which has a high incidence of myrmecochory. (Photo courtesy B. M. Campbell.)

Australia, possibly owing to slightly higher concentrations of phosphorus in the parent rock (Marchant & Moore 1978). We also chose two further sites, the fertile sites, less than 5 km away and experiencing similar climates, to compare the effects of soil fertility on dispersal strategies. The fertile sites resembled the infertile sites in soil texture and free drainage, but were situated on the vegetated lee of coastal sand dunes composed partly of calcium carbonate and containing relatively high nutrient concentrations. Again, actual values were higher in the South African than in the Australian site. In particular, potassium concentrations in calcareous sands were five times those in siliceous sands at the Caledon coast (Table 1). The sites are termed 'fertile' in a relative sense only.

Methods

Ant populations were sampled using test tubes sunk into the soil approximately 2 m apart in transects crossing the study sites, supplemented by hand collections and sight records. In each site at least 50 tubes were left functioning as pitfall traps for a continuous period of 5 days, and sampling was carried out in both late summer/autumn and late winter/spring. All plant species in the study sites were recorded. Characteristics of plant diaspores and mode of dispersal were observed in the study sites or, for species not fruiting at the time, inferred from the same or closely related species observed elsewhere.

Results

The plant and animal communities

The infertile site at the Barrens bears a heath of dwarf shrubs (mainly Proteaceae and Myrtaceae) and sedges (Cyperaceae and Restionaceae) with emergent *Banksia*. The infertile site at the Caledon coast bears structurally similar vegetation containing *Protea*, Ericaceae and many Restionaceae. The floras of the fertile sites are different. In Australia, shrubby *Eucalyptus, Melaleuca* and *Acacia* species were dominant, associated with semi-succulent herbs or low shrubs such as Chenopodiaceae, Aizoaceae, Sterculiaceae, Rhamnaceae, Asteraceae and Zygophyllaceae. In South Africa the main shrubs are *Euclea, Rhus, Sideroxylon, Cassine* and *Olea* species, and form dense clumps bound with vines such as *Cynanchum, Kedrostis* and *Asparagus,* with a seasonal herb layer. Floristic similarities are summarized in Table 2.

The ant communities in our study sites share some genera and sub-families but no species between continents (Table 3). Numbers of both species and individuals, as assessed by our sampling methods, are higher in the Australian than the South African sites, although further sampling is required. The extra numbers are made up largely of genera such as

Table 2. Floristic links between study sites, between continents (top) and within continents (bottom). Values in the top section are percentages of total taxa in each study site also found in the study site of similar fertility on the other continent, excluding species introduced by man from one continent to the other. Values in the lower section are percentages of total taxa in each study site also found in the study site of different fertility on the same continent. Restionaceae were regarded as belonging to separate genera on the two continents pending revision of the family.

	Families		Genera		Species	
	Aust.	S.Afr.	Aust.	S.Afr.	Aust.	S.Afr.
Plants shared between infertile sites	47	42	4	5	0	0
Plants shared between fertile sites	31	32	13	15	0	0

	Australia			South Africa		
	Families	Genera	Species	Families	Genera	Species
Infertile site	44	13	0	70	14	2
Fertile site	29	15	0	44	17	2

Table 3. Relative richness of ant communities in the study sites at the Barrens, Western Australia, and the Caledon coast, South Africa.

	Australia		South Africa	
	Infertile site	Fertile site	Infertile site	Fertile site
Total individuals trapped	803	1100	364	144
Number of species recorded	30	16	17	13
Number of genera recorded	19	13	10	10
% genera shared with other continent	31	38	50	56
Number of trap days	500	750	750	375

Rhytidoponera and *Iridomyrmex,* which are distinctively Australian (Berg 1975; Greenslade 1976, 1979; Majer, this volume). The genera *Pheidole, Monomorium, Tetramorium, Meranoplus, Crematogaster* and *Camponotus* are present on both continents. These form the major part of the South African community, together with *Anoplolepis steingroeveri,* not recorded in Australia.

Incidence of myrmecochory

The incidence of myrmecochory at the study sites is summarized in Table 4. Interactions observed in the Australian site agree with those described by Berg (1975, 1981), Clifford & Drake (1981), Davidson & Morton (1981, 1982), Drake (1981), Rice & Westoby (1981), Majer (this volume) and Westoby et al. (this volume), and the present discussion is therefore confined to the South African sites.

Detailed observations and experiments in the infertile study site, and on similar soils further afield (Slingsby & Bond 1981), show that the diaspores of many Cape fynbos plants are highly attractive to the local ants (Table 4). Small food bodies (elaiosomes) were identified for 29% of the plant species and 44% of the families at this site at the Caledon coast. There are several important similarities with Australia. Firstly, the myrmecochorous species are mainly long-lived woody or sclerophyllous perennials. Secondly, diplochory, with explosive release of seed with elaiosomes from dehiscent capsules or pods, is common, particularly in the Fabaceae, Rutaceae & Rhamnaceae. Thirdly, the myrmecochores in South Africa contribute only a small proportion of total plant cover or biomass. Fourthly, ants collect diaspores in South Africa in a similar manner to that described from Australia (Berg 1975; Drake 1981): the diaspores are discovered and removed very quickly, particularly the relatively large proteaceous

diaspores, and are often buried permanently in the ant nests. In myrmecochorous South African Restionaceae, the diaspores are unattractive to ants until ripe and released naturally by the plant.

Anoplolepis steingroeveri is by far the most important diaspore-collecting ant at the Caledon coast, together with *Camponotus niveosetosus.* The harvesters *Messor, Pheidole* and *Myrmicaria* have also been observed gathering seed with elaiosomes in ı Cape fynbos, but they are usually chased away by *Anoplolepis.* As in Australia (Drake 1981; Rice & Westoby 1981; Davidson & Morton 1982) and in North America (Culver & Beattie 1978; Heithaus et al. 1980), elaiosomes are not the ants' sole food source. Several of the elaiosome gatherers recorded from the infertile sites also occur in fertile sites in association with aphids and coccids (Prins 1978; Dr J. C. Taylor, personal communication). In summary, the number of plant species with elaiosomes in the infertile Australian site is not significantly greater than that in the infertile South African site, but the number of ant species which collect the elaiosomes is greater in Australia.

In striking contrast to the infertile site, myrmecochory is rare in the fertile South African site, being restricted to a few Euphorbiaceae and Zygophyllaceae (Table 4). Instead, the dominant plants fleshy fruits dispersed by birds. In the corresponding site in Australia, which is less fertile in absolute terms, a number of plants have fleshy fruits but myrmecochory is still fairly common. The birds at the Caledon coast include several species relying mainly on fleshy fruits for food, particularly colies (Coliidae) and bulbuls (Pycnonotidae) (Phillips 1926; Ridley 1930; Rowan 1970; McLachlan & Liversidge 1972; Palmer & Pitman 1972; Palgrave 1977) while the relatively few species known to eat fleshy fruits at the Barrens take a wide variety of other items as well (Noble 1975; Clifford & Drake 1981; Schodde 1981).

Table 4. Numbers and proportions of plant species dispersed by ants or other agents in each study site.

| Main dispersal agent | Australia | | | | South Africa | | | |
| | Infertile site | | Fertile site | | Infertile site | | Fertile site | |
	No.	%	No.	%	No.	%	No.	%
Ants	16	25	10	24	18	29	5	13
Birds and other vertebrates	1	1	13	32	0	0	14	35
Other, or unknown	49	74	18	44	44	71	21	52

Food bodies intermediate in size between fleshy fruits and elaiosomes, such as the small fruits of Chenopodiaceae or the large, bright arils of *Acacia* species, are probably taken by both birds and ants. Such structures are typical in the fertile site in Australia and do also occur in South Africa (e.g. *Pterocelastrus*). There are no fleshy-fruited plants in the infertile site on either continent, save the parasitic twiner *Cassytha* (Table 4).

Discussion

This study shows that myrmecochores are common where environmental conditions are suitable for them in southern Africa, providing a striking case of convergence with Australia. Dispersal strategies are more similar in like environments on different continents than in unlike environments within the same continent. Some of the myrmecochorous plants in South Africa are unrelated to those in Australia, while others belong to the same families. These floristic links do not detract from convergence, since the floras of nearby sites with different environments are quite distinct. There are similar patterns in the ant faunas, although these share a substantial number of genera between continents and differ less between sites than do the plants, presumably owing to the plasticity of their feeding habits. No species-specific interactions between ants and plants, or for that matter birds and plants, are known from either area.

The significance of convergence

Our results support the hypothesis that climate and substrate are important determinants of dispersal strategies. Most studies of convergence have concentrated on areas of similar climate following the early work of plant geographers (Schimper 1903; Diels 1906). More recently the fundamental role of soil nutrients in explaining non-convergence has been emphasized in studies of vegetation (Specht 1969, 1979; Cowling & Campbell 1980) and animal communities (Milewski 1981). Neither a mediterranean climate, shrub dominance, nor a regime of occasional fires is sufficient in itself to select for large numbers of woody species dispersed by ants. In garrigue growing on relatively fertile soils in the south of France (Specht 1969), myrmecochory is restricted to only 10 species, or 3.6% of the flora (Sernander 1906;

Müller 1933). The California chaparral is apparently also poor in myrmecochores (Berg 1966; Bullock 1974). However the syndrome extends well beyond the mediterranean climate in Australia. Myrmecochores occur on leached, siliceous sands under a mainly summer rainfall (Drake 1981), and also on phosphorus-poor or sodic soils under a semi-arid climate (Davidson & Morton 1981). They are also relatively common in European heathlands away from the Mediterranean Basin on acid, sandy soils, particularly in Fabaceae and Polygalaceae (Van der Pijl 1969).

This study suggests that fleshy fruits are restricted to plants with plentiful supplies of several nutrients, particularly potassium, whether obtained directly from the soil or by parasitizing other plants. Diaspores taken by birds possibly contain more nutrients than those taken by ants (Morton 1973; Rice & Westoby 1981; Davidson & Morton 1982). Although our study sites represent the extremes of soil fertility available at both the Barrens and the Caledon coast, the fertile site at the Barrens is less fertile than that at the Caledon coast. The higher incidence of ornithochory at the latter may therefore be due to better nutrient status, and perhaps also to the more frequent mists (Nagel 1962; Bureau of Meteorology 1971). Certain habitats outside the mediterranean zone in Australia, even under fairly dry climates, carry floras containing many ornithochores and few myrmecochores (Beadle 1966; Johnstone & Smith 1981; Clifford & Drake 1981; Rice & Westoby 1981). Further research is needed on the growth and ripening of large fleshy fruits and arils in relation to the uptake of phosphorus, calcium, potassium, nitrogen and trace elements, and the effects of high sodium or magnesium concentrations on this process. As a working hypothesis we suggest that ornithochory occurs only on soils containing more than 150 ppm total phosphorus and 0.15 milli-equivalents per 100 g exchangeable potassium, being replaced by myrmecochory if the concentration of either element is below this threshold.

Our data help to eliminate some other hypotheses for the evolution of myrmecochory in Australia (Berg 1975, 1981; Westoby et al., this volume). Firstly, the number of myrmecochores in Australia is probably not dependent on the richness of the Australian ant fauna, since the ant fauna of the Cape fynbos in South Africa is rich in neither species nor individuals (Prins 1978; this study). Secondly, unique historical events are an insufficient explanation since the two continents differ markedly in topography and thus the availability of climatic refugia, and in the past distribution patterns and migrations of the flora (Axelrod & Raven 1978; Deacon 1979; Barlow 1981). Thirdly, predation on seed by rodents (Heithaus 1981) cannot on its own account for myrmecochory. Rodents are very scarce in mediterranean Australia and are certainly more common at the Caledon coast, including the fertile soils, than at the Barrens (Milewski, unpublished data). It seems more likely that the scarcity of rodents facilitates myrmecochory, as suggested by Davidson and Morton (1981). Fourthly, convergence has occurred despite the possible involvement of co-adaptation between animals and plants, which might be expected to lead to non-convergence irrespective of the physical environment (Milewski 1982).

Why myrmecochory?

Strategies for seed dispersal by vertebrates appear to be unsuitable on nutrient-poor soils under a mediterranean climate for a number of reasons besides the plant's need for nutrients to produce the food body. Seeds occurring in fleshy fruits dispersed by birds tend to be short-lived (Bullock 1974). This, together with inefficient burial, would adapt them poorly to the fires which typically occur once every 1–2 decades in the very sclerophyllous, fine-leafed, resinous vegetation on these infertile soils (Beadle 1966; Gill 1975). Dispersal by fleshy fruits is instead suited to sites where nutrients are adequate for regeneration at any time between fires, and where the establishment of seedlings is restricted mainly by the availability of suitable gaps in the existing cover (Bullock 1978). Plants with fleshy fruits are generally concentrated in small areas protected from fire, and in the absence of fire, patches gradually coalesce to form more extensive stands that do not burn (Taylor 1961; Hnatiuk & Kenneally 1981). At the Caledon coast, the regenerating vegetation on the scorched margins of such patches comprises *Leucadendron, Passerina* and *Thamnochortus,* which form the mature vegetation on less fertile soils. This successional stage contains fewer ornithochores and more myrmecochores than the eventual climax which returns with the build-up of organic matter and nutrients in the soil (Taylor 1961).

Soil seed reserves are essential to plant survival in fire-prone vegetation. These may be maintained by modest production of long-lived seed or by frequent abundant production of short-lived seed. On infertile soils where nutrients limit plant growth, frequent production of abundant large seed is not feasible. In such situations, however, small seed may require mycorrhizal infection for successful germination and establishment (Harper 1977). The chances of a propagule reaching a suitable site for immediate establishment would be particularly small on infertile soils covered with long-lived, slow-growing plants. Under such conditions, dispersal to a favourable microenvironment is perhaps more valuable than dispersal far from the parent. Wind and birds can both disperse seed effectively over long distances, but wind-dispersed seed must be small, and bird-dispersed seed require fleshy fruit or arils, so both would be at a disadvantage in fire-prone vegetation on infertile soils. Dispersal by ants achieves prompt seed burial and thus escape from fire and predation, together with the benefits of regeneration in locally enriched ant nests or middens in a nutrient-poor environment. A hard protective seed coat ensures long seed viability and complements myrmecochory well, preventing seed damage whilst the diaspore is being carried to the ant nest (Berg 1975; Drake 1981; Westoby et al., this volume).

We therefore conclude that soil infertility, in conjunction with the mediterranean climate, imposes constraints on plant dispersal strategies – both by direct limitation of resources for diaspore production, and indirectly, by its effects on vegetation structure and consequent fire regime. This study supports Davidson & Morton (1982) on the importance of soil nutrients in the development of myrmecochory. Further comparisons of different continents, where biological variation can be tested against independent physical variables in a natural experimental design, could provide a better understanding of the ecology and evolution of interactions between animals and plants.

Acknowledgements

We are indebted to Dr J. D. Majer, without whose continuing help and support this study would not have been possible. We thank Dr D. W. Davidson, Dr F. J. Hingston, Mr P. Slingsby and Dr J. C. Taylor for assistance and discussion. However, the views expressed remain the responsibility of the authors. Mr J. G. Penniket and Dr A. J. Prins identified ant specimens and Mr H. Demarz, Mr G. J. Keighery and Mr K. R. Newbey identified plant specimens or elaiosomes. Assistance in collecting ants was kindly provided by Ms J. M. Waldock, Ms M. N. Sartori and Mr P. A. Towers in Australia and by Dr R. D. Wooller and Dr A. J. Williams in South Africa. Ms V. L. Hamley uncomplainingly retyped numerous drafts of the manuscript. This work is based on part of a Ph. D. thesis (A.V.M.) supervised by Dr R. D. Wooller at Murdoch University, while support for W.J.B. was provided by the Directorate of Forestry and Environmental Conservation, South Africa.

References

Aplin, T. E. H. & Newbey, K. R., 1982. A botanical survey of the Fitzgerald River National Park. West. Aust. Herb. Res. Notes, in preparation.

Axelrod, D. I. & Raven, P. H., 1978. Late Cretaceous and Tertiary vegetation history of Africa. Monographiae Biologicae 31: 77–130.

Barlow, B. A., 1981. The Australian flora: its origin and evolution. In: Flora of Australia, Vol. 1 (Introduction). Bureau of Flora and Fauna, Australian Government Publishing Service, Canberra.

Beadle, N. C. W., 1966. Soil phosphate and its role in molding segments of the Australian flora and vegetation, with special reference to xeromorphy and sclerophylly. Ecology 47: 991–1007.

Beard, J. S., 1981. The vegetation of the Swan area. Vegetation Survey of Western Australia, 1:1,000,000 vegetation series. University of Western Australia Press, Nedlands.

Berg, R. Y., 1966. Seed dispersal of Dendromecon: its ecologic, evolutionary, and taxonomic significance. Am. J. Bot. 53: 61–73.

Berg, R. Y., 1975. Myrmecochorous plants in Australia and their dispersal by ants. Aust. J. Bot. 23: 475–508.

Berg, R. Y., 1981. The role of ants in seed dispersal in Australian lowland heathland. In: Specht, R. L., (ed.), Heathlands and Related Shrublands: Analytical Studies, 41–50. Elsevier, Amsterdam.

Bullock, S. H., 1974. Seed dispersal of Dendromecon by the seed predator Pogonomyrmex. Madrono 22: 378–379.

Bullock, S. H., 1978. Plant abundance and distribution in relation to types of seed dispersal in chaparral. Madrono 25: 104–105.

Bureau of Meteorology (Australia), 1971. Climatic survey. Region 5: Esperance, Western Australia. Department of the Interior; Government Printer, Canberra.

Clifford, H. T. & Drake, W. E., 1981. Pollination and dispersal in eastern Australian heathlands. In: Specht, R. L. (ed.), Heathlands and Related Shrublands: Analytical Studies, 51–60. Elsevier, Amsterdam.

Cody, M. L. & Mooney, H. A., 1978. Convergence versus nonconvergence in mediterranean-climate ecosystems. Ann. Rev. Ecol. Syst. 9: 265–321.

Cowling, R. M. & Campbell, B. M., 1980. Convergence in vegetation structure in the mediterranean communities of California, Chile and South Africa. Vegetatio 43: 191–197.

Culver, D. C. & Beattie, A. J., 1978. Myrmecochory in Viola: dynamics of seed–ant interactions in some West Virginia species. J. Ecol. 66: 53–72.

Danserau, P. & Lems, K., 1957. The grading of dispersal types in plant communities and their ecological significance. Contrib. Inst. Bot. Univ. Montreal 71.

Davidson, D. W. & Morton, S. R., 1981. Myrmecochory in some plants (F. Chenopodiaceae) of the Australian arid zone. Oecologia 50: 357–366.

Davidson, D. W. & Morton, S. R., 1982. Coevolution of Acacia with ants and birds of the Australian arid zone. Ecology, in press.

Deacon, H. H., 1979. Palaeoecology. In: Day, J. & Siegfried, W. R., (eds.), Fynbos Ecology, a Preliminary Synthesis. S. Afri. Nat. Sci. Prog. Rep. 40.

Diels, L., 1906. The Plant World of Western Australia South of the Tropics. English translation by W. J. Dakin, 1920. Unpublished manuscript, Western Australian Herbarium, Perth.

Drake, W. E., 1981. Ant–seed interaction in dry sclerophyll forest on North Stradbroke Island, Queensland. Aust. J. Bot. 29: 293–310.

Gill, A. M., 1975. Fire and the Australian flora: a review. Aust. For. 38: 4–25.

Goldblatt, P., 1978. An analysis of the flora of southern Africa: its characteristics, relationships and origins. Ann. Miss. Bot. Gard. 65: 369–436.

Good, R., 1974. The Geography of the Flowering Plants. Longmans Green, London.

Greenslade, P. J. M., 1976. The meat ant Iridomyrmex purpureus (Hymenoptera: Formicidae) as a dominant member of ant communities. J. Aust. Ent. Soc. 15: 237–240.

Greenslade, P. J. M., 1979. A Guide to Ants of South Australia. South Australian Museum, Special Educ. Bull. Series, Adelaide, 44 pp.

Harper, J. L., 1977. Population Biology of Plants. Academic Press, London.

Harper, J. L., Lovell, P. H. & Moore, K. G., 1970. The shapes and sizes of seeds. Ann. Rev. Ecol. Syst. 1: 327–356.

Heithaus, E. R., 1981. Seed predation by rodents on three ant-dispersed plants. Ecology 62: 136–145.

Heithaus, E. R., Culver, D. L. & Beattie, A. J., 1980. Models of some ant–plant mutualisms. Am. Nat. 116: 347–361.

Hnatiuk, R. J. & Kenneally, K. F., 1981. A survey of the vegetation and flora of Mitchell Plateau, Kimberley, Western Australia. Biological Survey of Mitchell Plateau and Admiralty Gulf, Kimberley, Western Australia. Western Australian Museum, Perth.

Hopper, S. D., 1979. Biogeographic aspects of speciation in the south-west Australian flora. Ann. Rev. Ecol. Syst. 10: 399–422.

Janzen, D. H., 1969. Seed-eaters versus seed size, number, toxicity and dispersal. Evolution 23: 1–27.

Johnstone, R. E. & Smith, L. A., 1981. Birds of Mitchell Plateau and adjacent coasts and lowlands, Kimberley, Western Australia. Biological Survey of Mitchell Plateau and Admiralty Gulf, Kimberley, Western Australia. Western Australian Museum, Perth.

Marchant, J. W. & Moore, A. E., 1978. Geochemistry of the Table Mountain Group, II: analysis of two suites of western Graaf-water rocks. Trans. Geol. Soc. S. Afr. 81: 353–357.

McLachlan, G. R. & Liversidge, R., 1972. Roberts' Birds of South Africa. Trustees of the John Voelcker Bird Book Fund.

Milewski, A. V., 1979. A climatic basis for the study of convergence of vegetation structure in mediterranean Australia and southern Africa. J. Biogeog. 6: 293–299.

Milewski, A. V., 1981. A comparison of reptile communities in relation to soil fertility in the mediterranean and adjacent arid parts of Australia and southern Africa. J. Biogeog. 8: 493–503.

Milewski, A. V., 1982. Some differences between the plant and animal communities in mediterranean-type environments at the Barrens and the Caledon coast, as a basis for studies of evolutionary convergence in Australia and southern Africa.

Moll, E. J., McKenzie, B. & McLachlan, D., 1980. A possible explanation for the lack of trees in the fynbos, Cape Province, South Africa. Biol. Conserv. 17: 221–228.

Mooney, H. A., (ed.), 1977. Convergent Evolution in Chile and California: Mediterranean Climate Ecosystems. US/IBP Synthesis Series 5. Dowden, Hutchinson & Ross, Stroudsberg, Pennsylvania.

Mooney, H. A. & Dunn, E. L., 1970. Convergent evolution of mediterranean-climate sclerophyll shrubs. Evolution 24: 292–303.

Morton, E. S., 1973. On the evolutionary advantages and disadvantages of fruit eating in tropical birds. Am. Nat. 107: 8–22.

Müller, P., 1933. Verbreitungsbiologie der Gariqueflora. Beih. Bot. Zentralbl. 50: 395–469.

Nagel, J. F., 1962. Fog precipitation measurements of Africa's southwest coast. Notos 11: 51–60.

Noble, J. C., 1975. The effects of emus (Dromaius novaehollandiae Latham) on the distribution of the nitre bush (Nitraria billardieri DC). J. Ecol. 63: 979–984.

Orians, G. H. & Solbrig, O. T., 1977. Convergent Evolution in Warm Deserts. Dowden, Hutchinson & Ross, Stroudsburg, Pennsylvania.

Palgrave, K. C., 1977. Trees of Southern Africa. C. Struik, Cape Town.

Palmer, E. & Pitman, N., 1972. Trees of Southern Africa. 3vv. Balkema, Cape Town.

Parsons, D. J., 1976. Vegetation structure in the mediterranean scrub communities of California and Chile. J. Ecol. 64: 435–447.

Parsons, D. J. & Moldenke, A. R., 1975. Convergence in vegetation structure along analogous climatic gradients in the mediterranean climate ecosystems of California and Chile. Ecology 56: 950–957.

Phillips, J. F. V., 1926. General biology of the flowers, fruits and young regeneration of the more important species of the Knysna Forest. S. Afr. J. Sci. 23: 366–417.

Pianka, E. R., 1979. Diversity and niche structure in desert communities. In: Goodall, D. W. & Perry, R. A., (eds.), Arid-Land Ecosystems: Structure, Functioning and Management, Vol. 1, 321–341. Cambridge University Press, Cambridge.

Prins, A. J., 1978. Hymenoptera. In: Werger, M. J. A. (ed.), Biogeography and Ecology of Southern Africa, 823–875. Junk, The Hague.

Raven, P. H. & Axelrod, D. I., 1974. Angiosperm biogeography and past continental movements. Ann. Miss. Bot. Gard. 61: 539–673.

Rice, B. L. & Westoby, M., 1981. Myrmecochory in sclerophyll vegetation of the West Head, New South Wales. Aust. J. Ecol. 6: 291–298.

Ridley, H. N., 1930. The Dispersal of Plants Throughout the World. Reeve, Ashford, Kent.

Rowan, M. K., 1970. The foods of South African birds. Ostrich, suppl. 8: 343–356.

Salisbury, E. J., 1942. The Reproductive Capacity of Plants. Bell, London.

Schimper, A. F. W., 1903. Plant Geography on a Physiological Basis. Claredon Press, Oxford.

Schodde, R., 1981. Bird communities of the Australian mallee: composition, derivation, distribution, structure and seasonal cycles. In: Di Castri, F., Goodall, D. W. & Specht, R. L. (eds.), Mediterranean-Type Shrublands, 387–415. Elsevier, Amsterdam.

Sernander, R., 1906. Entwurf einer Monographie der europaischen Myrmekochoren. Sv. Vet. Ak. Handl. 41: 1–410.

Slingsby, P. & Bond, W. J., 1981. Ants – friends of the fynbos. Veld and Flora 67: 39–45.

Specht, R. L., 1969. A comparison of the sclerophyllous vegetation characteristic of mediterranean type climates in France, California, and southern Australia. II. Dry matter, energy and nutrient accumulation. Aust. J. Bot. 17: 293–308.

Specht, R. L. (ed.), 1979. Heathlands and Related Shrublands: Descriptive Studies. Elsevier, Amsterdam.

Spies, J. J., Engelbrecht, L. N. J., Malherbe, S. J. & Viljoen, J. J., 1963. Die geologie van die gebied tussen Bredasdorp en Gansbaai. Department of Mines Geological Survey, South Africa. Government Printer, Pretoria.

Stebbins, G. L., 1971. Adaptive radiation of reproductive characteristics in angiosperms, II: Seeds and seedlings. Ann. Rev. Ecol., Syst. 2: 237–260.

Taylor, H. C., 1961. Ecological account of a remnant coastal forest near Stanford, Cape Province. J. S. Afr. Bot. 27: 153–165.

Taylor, H. C., 1978. Capensis. In: Werger, M. J. A. (ed.), Biogeography and Ecology of Southern Africa, 171–229. Junk, The Hague.

Van der Pijl, L., 1969. Principles of Dispersal in Higher Plants. Springer, Berlin, 154pp.

Walsh, B. N., 1968. Some notes on the incidence and control of driftsands along the Caledon, Bredasdorp and Riversdale coastline of South Africa. Department of Forestry (South Africa) Bulletin, 44. Government Printer, Pretoria.

Walter, H., 1979. Vegetation of the Earth and Ecological Systems of the Geo-Biosphere. 2nd. edn. (transl. J. Wieser). Springer, New York.

CHAPTER TEN

Evidence for interspecific competition influencing ant species diversity in a regenerating heathland

Marilyn D. Fox and Barry J. Fox

Abstract. Ant species diversity was monitored along a mining path rehabilitated four to eleven years previously. The mining path at Hawks Nest ($32°30'$S, $152°30'$E) passes through closed heath and scrub with some open forest patches. Ant species richness, ant species diversity and the number of individuals seemed to be more dependent on the structure of the ant community present than on site variables. The equitability component however showed a linear increase with time, from which it could be predicted that mined plots would not reach the values on control plots in less than 17 years (95% C.I. 13–21 yrs).

An abrupt replacement of species comprising the ant community occurred approximately nine years after rehabilitation. This can be related to replacement of the dominant species and has been interpreted as interspecific competition determining the structure and hence the diversity of the ant community.

Predictive multiple regression equations for species diversity, species richness and the number of individual ants, on mined plots, account for 73%, 79% and 93% of each variance respectively. Independent variables used in the equations were: a floristic component (the absence of a group of plant species found in mature heath); foliage height diversity; the amount of vegetation in vegetation layers (20 to 50 cm and 50 to 100 cm); and soil hardness.

Introduction

A regenerating sand mining path provides a range of simplified habitats, from a very sparse vegetation cover, progressing towards the vegetation cover of the surrounding plant community. It enables changes in environmental variables to be tracked through time, to reveal their regeneration response. The response of the ant community can also be followed, in terms of diversity and abundance parameters, as functions of regeneration time. The relationship between ant parameters and environmental variables can readily be examined over the range of values available, to see which factors may be influencing ant community changes. We use the Hawks Nest mining path to attempt to answer these questions. A substantial amount of information about the Hawks Nest mining path is already available (Fox 1978; Fox & Fox 1978), and the sampling technique we use has been well examined (Green-

slade & Greenslade 1971; Greenslade 1973).

Several studies of ant recolonization are reported in the literature: recolonization on a series of areas rehabilitated after bauxite mining in Western Australia, shows 89% of the species returning within 13 years (Majer 1978; Majer et al., in press); and disturbance caused a marked reduction in the species diversity of a tropical ant fauna (Greenslade & Greenslade 1977); while Brian et al. (1976) have reported changes in species distributions in heath regenerating after fire. The mosaic nature of spatial patterning in ant communities has been well documented in English heathland (Brian 1964; Brian et al. 1965), and in forest in Ghana and Papua New Guinea (Majer 1972; Room 1975a). A similar pattern was found for an ant community in the Australian semi-arid zone (Briese & Macauley 1977, 1980). A number of authors have related this spatial patterning or the resulting species diversity to interspecific competition in the ant communities (Haskins & Haskins 1965; Crowell

Buckley, R. C. (ed.), Ant-plant interactions in Australia.
© 1982, Dr W. Junk Publishers, The Hague. ISBN 90 6193 684 5.

1968; Greenslade 1971; Lieberburg et al. 1975; Room 1975b), the existence of such competition having previously been verified experimentally by Pontin (1961, 1963).

The study area

Hawks Nest (32°30'S, 152°30'E) is located immediately north of Port Stephens. The mining path studied lies north of Hawks Nest on the outer barrier system of the Fens embayment (Thom 1965), paralleling the coast, inland from the hind dunes, but east of the Lower Myall River (Fig. 1).

The outer barrier system supports a tall dry heath, ranging from open heath to closed scrub (Specht et al. 1974). The heath is dominated by *Banksia, Casua-*

rina, Leptospermum and *Melaleuca* shrubs. Patches of open forest, dominated by *Eucalyptus pilularis* and *Banksia serrata,* occur on slightly higher ground. The mining path passes through this heath and occasional small open forest patches, whilst some sectors adjoin a swamp forest community dominated by *Eucalyptus robusta.*

Study plots established on the mining path represent a time sequence, as mining began at the southern end in 1965, proceeding north to finish at HN7 in 1972 (Fig. 1). Experimental plots were established on areas of mining path for which top soil was replaced in: April 1966 (HN1); October 1967 (HN3); January 1969 (HN4); February 1971 (HN5); November 1971 (HN6); and March 1972 (HN7). Control areas (HN2C and HN7C) were set up in heath adjacent to the mining path while HN8 was a control plot in

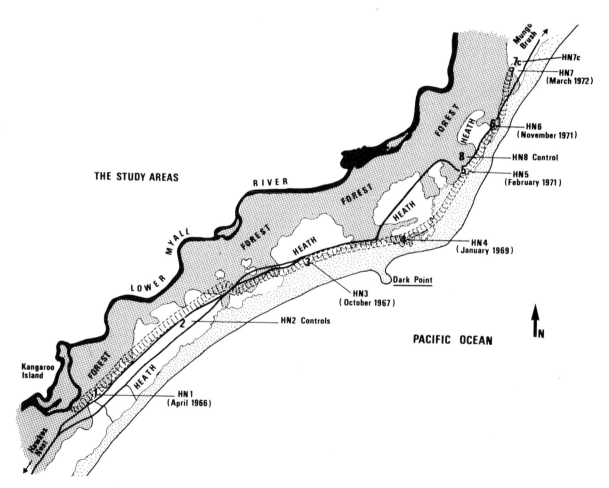

Fig. 1. Schematic representation of the Hawks Nest mining path with positions of the study plots and distribution of heath and forest; dates are for topsoil replacement. [After Fox & Fox (1978), reproduced with permission from Aust. J. Ecol.]

nearby forest. HN2B (burned in December 1972) was set up across a fire break from HN2C; HN7C was probably last burned in 1968, while HN2C had not burnt since at least 1960.

A complete vegetation survey and plant species list for the area, together with rehabilitation procedures for the experimental plots are given in Fox & Fox (1978).

Methods

Ant survey

At each experimental and control plot a 5 × 3 grid was established using a 20 m spacing between points. At each point a 2.5 × 10 cm specimen tube was sunk flush with the soil surface. Holes were augered with a bevelled 2.5 cm diameter steel tube. The pitfall traps contained 15 to 20 mls of 70% alcohol and a few drops of glycerol. Traps were set in place in the late afternoon, and picked up 24 hrs later (± 30 min), during a rain-free period. Four collections were made: April and December 1976; and April and December 1977. Ants were identified to genus using the key published in Greenslade (1979). A reference collection was prepared and used to allocate speciments to species, and individuals were counted. (Identifications were checked by R. W. Taylor and the reference collection deposited with A.N.I.C.)

Vegetation survey

Presence or absence in each of ten 1 m quadrats on each study plot provided a frequency measure for each plant species. Percentage cover values for individual species were measured as intercepts along a 100 m line transect. Vegetation height was measured every 3 m along this transect and the presence or absence of bare sand was recorded at 1 m intervals. Soil hardness was measured at twenty points with a penetrometer. Incident light readings at 1, 20, 50, 100, 150 and 200 cm above ground level were measured at forty points on each study plot, using a Gossen Lunasix light meter with incident light cone fitted. A vegetation index representing the amount of vegetation in each layer was calculated using the light flux immediately above and below the layer (see Fox 1979).

Results

Changes in vegetation structure as a function of regeneration time are shown in Fig. 2. HN7 shows an abnormally high value for the vegetation index in the 0 to 20 cm layer, caused by a thick mat of Brown's love grass, *Eragrostis brownii*. This caused the 0 to 20 cm vegetation index to be significantly skewed ($p < 0.05$) so it was excluded from analyses. Apart from HN7, the 0 to 20 cm layer increases up to eight years, after which it decreases to eleven years with control plot values even lower. Other layers show an overall increase with regeneration time, but with little vegetation above 100 cm until after eight years. A detailed examination of changes in vegetation structure and percentage cover changes in individual plant species is given in Fox & Fox (1978).

The total number of ant species found in all four collections is shown as a function of the mean regeneration time for each plot (from April 1976 to December 1977) (Fig. 3). The total number of individuals encountered on each study plot over the four collections is also shown (Fig. 3). The total number of ant species encountered is high after five years regeneration, low between five and eight years and high again from nine to eleven years. The total number of individuals encountered increases in the first eight years regeneration but decreases thereafter. The negative correlation between these two variables can be seen by graphing the mean number of

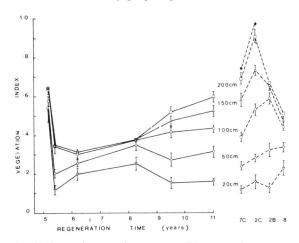

Fig. 2. Changes in vegetation structure with regeneration age on a sand mining path. The vegetation index is shown for each layer with ±1 s.e. shown at the top level of the layer. Control plot values are represented independent of the time scale. The identification of the symbols is given in the figure. [After Fox & Fox (1978), reproduced with permission from Aust. J. Ecol.].

individuals against the mean number of species per trapping session on mined plots (Fig. 4; $r = -0.85$, $p < 0.05$, $n = 6$). The control plots, however, do not follow this trend, having many fewer individuals for the same species richness.

The number of ant species encountered on each plot as a function of regeneration time is very similar for each collection (Fig. 5). This pattern suggests that the factors affecting the number of species present may be more closely related to environmental variables on each plot rather than regeneration time per se. However, ant species diversity, calculated as an equivalent number of equally abundant species (Hurlbert 1971; Simpson 1949), exhibits a signifi-

cant linear regression on regeneration time (Fig. 6; ant species diversity $= -0.67 + 0.38$ (regeneration age); $r = 0.65$, $p < 0.01$, $n = 24$). This relationship results from the increase in the equitability component of diversity with regeneration time (Fig. 7; equitability $= -0.017 + 0.022$ (regeneration age); $r = 0.65$, $p < 0.01$, $n = 24$). One sample from HN3 (April 1976) has a much higher equitability value than any other sample, perhaps due to the low number of individuals encountered. Without this point the correlation is markedly improved ($r = 0.75$, $p < 0.001$, $n = 23$).

A series of 30 environmental variables measured at these study sites were screened by Fox & Fox (1978) for inclusion as independent variables in a multiple linear regression analysis. In addition to these, we included five further independent variables, namely the scores of each plot on each of the first three components of a principal component analysis of the floristic data, the biomass of *Pseudomys novaehollandiae,* and the total rodent biomass *(Pseudomys +*

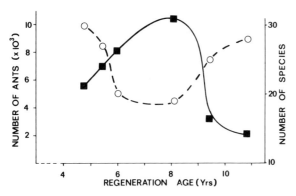

Fig. 3. The total number of species encountered on each plot (circles), and the total number of individual ants encountered on each plot (solid squares), both shown as functions of regeneration age.

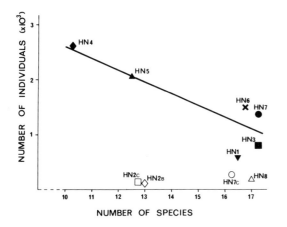

Fig. 4. The mean number of individuals per collection as a function of the mean number of species taken in each collection. Values are shown for both mined and control plots. The regression equation for all mined plots is $y = 4787 - 217x$, $r = 0.823$, $p < 0.05$, $n = 6$.

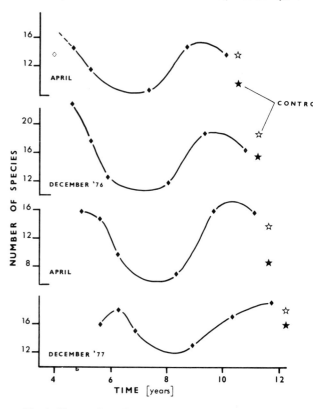

Fig. 5. The number of species encountered on each plot as a function of regeneration time. Values are shown for each of the four collections separately. Control values are included independent of the time scale: hollow stars, 7C; solid stars, 2C.

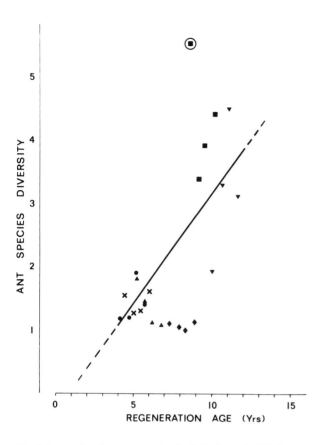

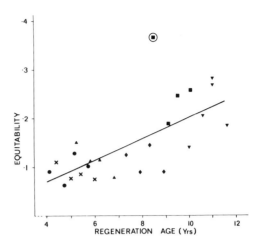

Fig. 7. The equitability component of species diversity on each mined plot for each of the four collections as a function of regeneration age. Symbols as for Fig. 4. The abnormal value for the April 1976 collection on HN3 is circled (see text). Equitability is calculated as $E = \Delta_3 / S$, where Δ_3 is ant species diversity and S is the number of species present.

Fig. 6. Ant species diversity on each mined plot for each of the four collections as a function of regeneration age. Symbols as for Fig. 4. The abnormal value for the April 1976 collection on HN3 is circled (see text for explanation). Ant species diversity (Δ_3) is calculated as $\Delta_3 = [\Sigma P_i^2]^{-1}$, where P_i is the proportion of the total number of individuals belonging to species i. Δ_3 is the equivalent number of equally abundant species with the same diversity (Simpson 1949).

Mus musculus). Where a group of variables were highly correlated (see Fox & Fox 1978, Fig. 4) only one variable was chosen from any group. The correlation matrix for the eleven independent variables chosen and the three dependent variables is shown in Table 1. Correlations with regeneration age are included in the table but this variable was not used in the multiple regression, as this would have required exclusion of other variables related to age.

December collections had a greater ant species richness than those in April (see Fig. 5). On the control plot HN7C, ant species were ranked separately on abundance at each of the four collections. These rankings were then tested using Spearman's rank correlation (Table 2). Both Aprils and both Decembers are highly significantly correlated ($p <$

0.001). For between seasons within a year, 1976 shows significant correlation ($p < 0.05$) but 1977 does not ($0.1 > p > 0.05$). Because of this, the multiple regressions were run separately for seasons (April and December), but results from both years were pooled.

Multiple regressions equations, significance levels and each variable's contribution to the variance in ant species diversity are shown in Table 3. All species that occurred in a minimum of two traps have been used. There is no significant difference between the April and December equations, although 73% of the variance is accounted for in December but only 54% in April. The December equation also has a higher significance level ($p < 0.01$). In view of this we will present equations for the number of species and the number of individuals for December only (Table 4). The proportion of the variance attributable to each variable is calculated from the contribution coefficient for that variable (see Fox & Fox 1978 for further explanation).

Discussion

Examination of the number of ant species encountered in at least two traps, and the number of individual ants (Fig. 3) indicates a distinct non-linearity, in

103

Table 1. Pairwise correlations between the eleven independent and three dependent variables.

X_n		1	2	3	4	5	6	7	8	9	10	11	12	13	14
Ant species richness	1														
Number of individuals	2	-0.511													
Ant species diversity	3	-0.372	-0.794												
Floristic component one	4	-0.399	0.752	-0.768											
Percent bare sand	5	-0.570	0.512	-0.262	0.230										
C.V. % light reaching ground	6	-0.100	0.021	0.281	-0.184	0.196									
Penetrability 0 to 15 cm	7	-0.097	0.182	0.213	-0.322	0.284	-0.010								
Number of plant species	8	-0.021	-0.316	0.458	-0.521	0.156	-0.050	0.331							
Vegetation height	9	0.422	-0.279	0.308	-0.568	-0.563	0.048	0.415	0.380						
Total vegetation index	10	0.150	-0.248	0.277	-0.139	-0.557	0.532	-0.163	0.017	0.491					
20 to 50 cm vegetation index	11	0.408	-0.186	0.427	-0.176	-0.448	-0.225	0.157	0.338	0.378	0.241				
50 to 100 cm vegetation index	12	0.791	-0.428	0.00	-0.218	-0.643	-0.207	-0.263	-0.406	0.316	0.155	0.00			
Foliage height diversity	13	0.459	-0.664	0.622	-0.361	-0.541	-0.061	0.077	-0.077	0.187	0.366	0.388	0.442		
Heathness	14	0.563	-0.629	0.616	-0.775	-0.502	0.309	0.083	0.519	0.800	0.540	0.302	0.316	0.233	
Regeneration age	15	-0.039	-0.489	0.793	-0.450	-0.200	0.102	0.191	0.658	0.263	0.323	0.584	-0.416	0.399	0.42

Table 2. Values for Spearman's rank correlation coefficient between abundance ranks ($n = 24$) for ant species collections from four different time periods.

Comparison	r_s	Probability
April 1976/ December 1976	0.410	$p < 0.05$
April 1977/ December 1977	0.367	$p < 0.1$
April 1976/ April 1977	0.675	$p < 0.001$
December 1976/ December 1977	0.748	$p < 0.001$

variables that might reasonably be expected to be linear functions of regeneration age. The mean number of ant species per trap as a function of regeneration age is markedly disjunct, with the four youngest plots decreasing (4.4, 4.1, 3.2, 2.8), followed by a substantial jump to the two oldest plots, which themselves again start to decrease (5.6, 4.7). A similar effect is also seen in Fig. 4 where the two oldest plots HN1 and HN3 appear more aligned with the control

plots, rather than the four younger mined plots.

As none of the environmental variables measured are similarly disjunct in time, the answer may lie within the ant communities. This was examined by analyzing the frequencies of each species at each collection on each plot, including controls, measured as the number of traps occupied out of a total of 15 on each plot. The analysis was performed on the CSIRO CYBER 76 at Canberra, using the TAXON program library (Williams 1976). The 34 cases were classified on the frequency data for a total of 41 ant species, using program MULCLAS (Williams 1976), followed by a principal co-ordinate analysis (Gower 1966) using program GOWER (Williams 1976), and the minimum spanning tree between them calculated with program MINSPAN. A graphical representation of the results of this process, referred to as a minimum spanning ordination (for further information see Gillison 1978), is given in Fig. 8. The figure shows clear separation between younger (HN7,

Table 3. Multiple regression equations for ant species diversity (Y_3) showing the proportion of the variance explained by floristic component 1 (X_4) and foliage height diversity (X_{13}). There is no significant difference between the regression coefficients for April and December for either variable (X_4, $t = 0.41$; X_{13}, $t = 0.36$) Significance levels are given for F-ratios for the equation (Y_3), ($F_{2,9}$) for each variable ($F_{1,10}$ and $F_{2,9}$).

April	Y^*_3	$1.675 - 0.399 (\pm 0.185)$	X^*_4	$+ 0.740 (\pm 0.501)$	X_{13}
Variance					
explained	$54\% =$		34%	$+$	20%
F-ratio	5.2		7.4		2.2
	$p < 0.05$		$p < 0.025$		$p < 0.1$
December	$Y_3 =$	$1.311 - 0.315 (\pm 0.094)$	X_4	$+ 0.540 (\pm 0.255)$	X_{13}
Variance					
explained	$73\% =$		48%	$+$	25%
F-ratio	12.0		14.4		4.5
	$p < 0.005$		$p < 0.005$		$p < 0.05$

Table 4. Multiple regression equations for the number of ant species (Y_1) occurring in at least two traps and the number of individual ants (Y_2) encountered from those species in the December collection. Symbols as for Table 3, with the additional variables: 50 to 100 cm vegetation index (X_{12}); 20 to 50 cm vegetation index (X_{11}); penetrability 0 to 15 cm (X_7).

	$Y_1 = 11.28 +$	$2.67 X_{12}$	$+ 1.97 X_{11}$	
Variance				
explained	$79.2\% =$	62.5%	$+ 16.7\%$	
F-ratio	17.1	16.8	7.2	
	$p < 0.001$	$p < 0.005$	$p < 0.025$	
	$Y_2 = 8.89 +$	$3.21 X_4$	$-5.05 X_{13}$	$+ 4.74 X_7$
Variance				
explained	$93\% =$	56%	$+ 29\%$	$+ 8\%$
F-ratio	33.9	13.0	9.4	12.5
	$p < 0.01$	$p < 0.005$	$p < 0.01$	$p < 0.005$

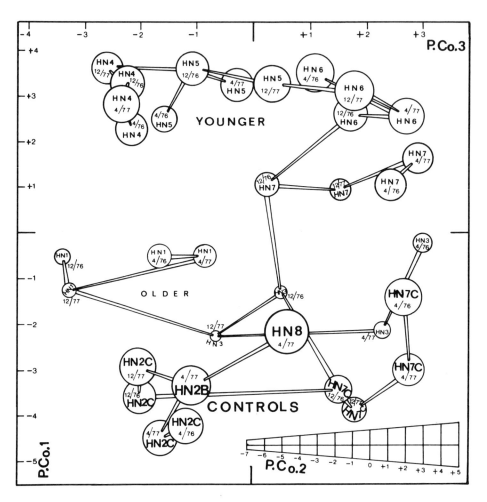

Fig. 8. A minimum spanning ordination (Gillison 1978) of ant communities from 34 separate collections: on six mined plots of increasing age, HN7, HN6, HN5, HN4, HN3 and HN1; and on 4 control plots HN7C, HN2C, HN2B and HN8 (see text). The date for each collection is shown on the figure. The points are displayed as the projection onto the plane of the first and third principal coordinate axes. The second principal co-ordinate axis is shown projecting perpendicular to the figure and is represented by different sized spheres as shown in the key, with larger spheres above and small spheres below the plane.

HN6, HN5 and HN4) and older plots (HN3 and HN1), which are in turn separable from the control plots. These separations are based solely on the ant community structure of each plot at each collection, as determined by the ant species present and their capture frequency.

The above data provide a plausible explanation for differences between plots, but give no indication either of the mechanism involved or of the reason for such abrupt changes in ant variables as a function of time. This information, however, can be derived from a closer examination of the ant communities themselves (Fig. 9). Five species of *Iridomyrmex*, and *Tapinoma minutum*, are numerically dominant

in these communities. Their relative abundances do not change gradually and uniformly with plot age, but switch suddenly between HN4 and HN3 from a community dominated by *Iridomyrmex* spp.C, B and D to one dominated by *Iridomyrmex* sp.A, *I. nitidiceps* and *T. minutum*. Not only are the dominant species replaced, but they carry with them entirely different communities, which not surprisingly produce markedly different values for the number of species, the number of individuals and species diversity, across the changeover point.

A close examination of the changes in frequency of the dominant species in each community (Fig. 10) shows how sharp the changeover can be, occurring

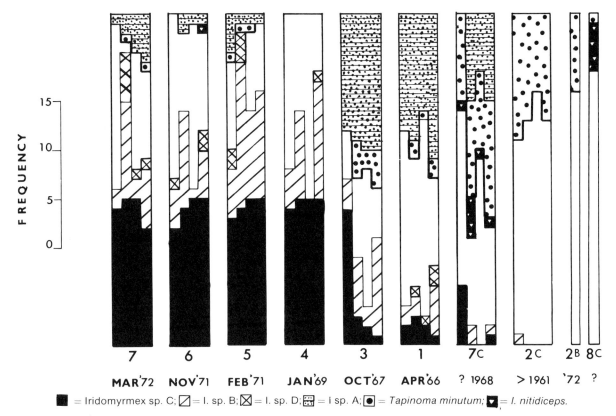

Fig. 9. Changes in relative importance of *Iridomyrmex* species as a function of time. The four collections for each plot are shown as small columns within each large column, arranged chronologically from left to right. Species associated mainly with the younger plots are arranged from the bottom of the figure while species associated mainly with the older plots are arranged from the top of the figure. A scale for the frequency of occurrence for each species (maximum 15 per plot), is shown at the left.

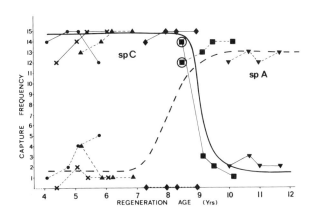

Fig. 10. Capture frequency for *Iridomyrmex* sp. A and *I.*sp.C for each collection on each mined plot as a function of regeneration time. Symbols are as for Fig. 4. Values for the April 1976 collection on HN3 are circled.

between the April and December collections in 1976. The reciprocal abundance exhibited by these two species shows a very strong negative correlation ($r = -0.90, n = 24, p < 0.001$). The results are not significantly different from the regression slope of -1.0 expected for a one-to-one replacement ($t = 0.952, 0.4 > p > 0.3$). This would be consistent with a high level of interspecific competition and is further supported by the lack of intermediate frequencies. With the exception of one sample, each plot is strongly dominated by either *Iridomyrmex* sp.C or *I.* sp.A. With both species abundant on HN3 in April 1976, at frequencies of 14 and 12 respectively from a maximum of 15, the site must be considered suitable for both; yet by December 1976 the frequencies were 3 and 13 respectively. This is consistent with competitive reduction of *Iridomyrmex* sp.C by *I.* sp.A; the evidence is inferential, but additional evidence is provided by the parameters for the April 1976 collection on HN3. These show marked departures from

107

values for other collections on HN3, as well as for all other plots (see circled point on Fig. 6, species diversity and Fig. 7, equitability). This is also borne out by the extremely low number of individuals in this collection: 292 as against 998, 834 and 1008 for the three other collections on HN3.

Further evidence derives from point richness values; the mean number of ant species caught per trap. Room (1975b) suggested that mean point richnesses represented the maximum number of overlapping ant territories, and cited values of 4.65 for a cocoa forest in Ghana and 4.57 for rain forest in Papua New Guinea. The value for our oldest mined plot (HN1) is 4.70, showing good agreement with these values, but that for plot HN3, where the changeover was occurring, was 5.60, exceeding Room's hypothesized maximum. This indicates a non-equilibrium situation where additional species are present as the community adjusts to the changeover.

All of these factors point to an abnormal occurrence at the time of the April 1976 collection on HN3. Interference competition between species that normally avoid one another would be such an occurrence. Such confrontation events between dominant species have been reported in the literature, though they are infrequent, with the species concerned actively avoiding one another for the majority of the time (Lieberburg, Kranz & Siep 1975). In this way interspecific competition can serve as the mechanism maintaining the partitioning of space that results in the mosaic distributions so often observed in ant communities (Brian et al. 1965; Haskins & Haskins 1965; Crowell 1968; Lieberburg et al. 1975). Briese & Macauley (1977) have shown mosaic spatial patterning in semi-arid ant communities in Australia, and conclude that community structure is largely determined by processes involving aggressive interactions between species.

Frequency changes associated with species replacement have been reported for *Anoplolepis longipes* and *Oecophylla smaragdina* in coconut plantations on the Solomon Islands (Greenslade 1971). The frequency changes closely followed a one-for-one exchange of palms between the species. The change in the dominant species also affected ant species diversity. When *A. longipes* became almost universally dominant, diversity fell abruptly; and as *A. longipes* declined, the diversity increased. The influence that a dominant species *(Iridomyrmex purpureus)* may have on community structure has been

well documented by Greenslade (1976). The instability present in the series of species replacements occurring in the structurally simple plantation environment probably resulted from the vegetation change over the 20-year period studied, and Greenslade (1971) concludes 'the ultimate process regulating diversity appears to be interspecific competition'.

This, and our own similar conclusion, remain inferential, being based on interpretations of past and present species' distributions. The existence of interspecific competition in ants is well documented; *Lasius flavus* colonies, for example, showed a significant increase in alate queen production when an adjacent *L. niger* colony was removed (Pontin 1963). Queen production is dependent upon adequate food being available and Pontin concluded that *L. niger* limited the food available to *L. flavus*. The reciprocal effect also occurred; when *L. flavus* colonies reached maturity, *L. niger* colonies were entirely excluded because they were denied suitable nest sites by the superior nest-building ability of *L. flavus* (Pontin 1963). *Lasius niger, L. alienus* and *Tetramorium caespitum* have been shown to maintain food territories in heathland in southern England, while *Formica fusca* and *Myrmica* species maintain nest territories. The result is a mosaic of blocks of heathland (food territories) in which only one species will be found (Brian et al. 1965). Detailed observations of fighting between *Pheidole megacephala* and *Iridomyrmex humilis* on Bermuda (Lieberburg et al. 1975) led to the conclusion that interference competition was taking place, with distributions becoming more of a mosaic than in previous surveys (Haskins & Haskins 1965; Crowell 1968). Lieberburg et al. suggest that an equilibrium distribution of the two species could be achieved. It would seem that such an equilibrium would be achieved and maintained by the effect each species has on the other; i.e., by interspecific competition.

In this study neither ant species richness nor the number of individuals encountered is a linear function of time (Fig. 3) so it would be very difficult to use either parameter to estimate the time required to reach control values. Ant species diversity (Fig. 6) has a statistically significant linear relationship, but it also shows a sharp break between HN4 and HN3, indicating that it is related more to the ant community composition than to regeneration time. Equitability increases more gradually with time, without disjunctions, and is therefore a more reliable predictive parameter. The mean value for the equitability component of diversity, based on 10 collections from control sites, is 0.350 ± 0.038 with a 95% confidence interval of 0.262-0.438. Assuming that equitability is a linear function of time, mined plots would require 16.7 years to reach this control mean value (95% confidence interval 12.7 to 20.7 years). The time required could be much longer if the equitability tends to overshoot before stabilizing at some lower plateau value, as indicated by the lower diversity of the oldest control plot, HN2C. Recovery times for total plant biomass, *Pseudomys* biomass and plant species diversity have been estimated at 20, 20 and 13-17 years respectively (Fox & Fox 1978). The above estimate for the regenerating ant community is in good agreement. For comparison, Majer et al. (in press), studying ant recolonization on rehabilitated bauxite mines, found that only five (11%) of a total community of 47 ant species had not recolonized by the twelfth or thirteenth year. This would be consistent with a recovery time of 15 to 20 years. Their multiple regression equation for ant species richness predicts recovery times of 15 years or 23 years, depending whether three or four variables are used. These results are all in agreement with our estimated recolonization rates. In a tropical ant fauna, however, only two thirds of the fauna had returned 20 years after disturbance of the rain forest (Greenslade & Greenslade 1977).

For both ant species diversity and the number of individuals on mined plots, the greatest proportion of the variance was accounted for by one of the floristic variables: scores on one of the components of the floristic PCA. The main characteristic of this component is the lack of a particular group of plant species, generally associated with the oldest study plots but not necessarily with the control plots. A large proportion of the variance was also contributed by foliage height diversity (FHD), an overall measure of the vegetation structure profile. A smaller additional contribution to the variance in the number of individuals was attributed to soil hardness, which probably affects nest building. The number of individuals is greatest when the FHD is least (i.e. fewest layers of vegetation), whereas the ant species diversity is greatest when the FHD, and hence the number of layers, is greatest. Ant species richness, however, is dependent on specific structural variables: almost two thirds of its variance is contributed by the vegetation index for the 50-100 cm (shrub)

vegetation layer and a smaller component by the 20-50 cm layer. The herb layer, 0-20 cm, was excluded from the analysis because of its non-normality, as noted earlier.

On sites undergoing rehabilitation after bauxite mining, the variables contributing significantly to ant species richness were plant species richness, time since start of rehabilitation, and percentage plant cover (Majer et al., in press). Those contributing significantly to overall ant species diversity were plant species diversity, plant cover, litter cover, and time since start of rehabilitation. For the Hawks Nest sites, plant species richness and regeneration age are correlated with each other ($p < 0.005$), and with the vegetation index for the 20-50 cm layer ($p < 0.05$; see Fox & Fox 1978, Fig. 4). Ants therefore appear to key on similar parameters in both studies. The factors determining ant parameters on control plots, however, may be different to those determined by the multiple regression equations for the mined plots, as the values predicted for the control plots are quite different to those actually measured. The reason for this is probably that the control plots support a markedly different ant community (see Fig. 8).

In conclusion, the factors influencing the diversity and abundance parameters of ant communities in regenerating heathland are complex, and may be determined by the structure of the ant community itself, as well as by site attributes. Changes in the site-controlled factors, as functions of time in this study, can trigger massive changes in ant community structure via the mechanism of interspecific competition.

Acknowledgements

We are glad to acknowledge the considerable assistance of John Greenslade and Bob Taylor, who both gave freely of their time and advice in identifying ants and discussing this project in its early stages. The computer analyses used were facilitated by the efforts of Andy Gillison, whom we thank. We are grateful for comments on an earlier draft of the manuscript from Ross Crozier, John Greenslade, Jonathan Majer and Bob Taylor.

References

Brian, M. V., 1964. Ant distribution in a southern English heath. J. Anim. Ecol. 33: 451–461.

Brian, M. V., Hibble, J. & Stradling, D. J., 1965. Ant pattern and density in a southern English heath. J. Anim. Ecol. 34: 545–555.

Brian, M. V., Mountford, M. D., Abbott, A. & Vincent, S., 1976. The changes in ant species distribution during ten years post-fire regeneration of a heath. J. Anim. Ecol. 45: 115–133.

Briese, D. T. & Macauley, B. J., 1977. Physical structure of an ant community in semi-arid Australia. Aust. J. Ecol. 2: 107–120.

Briese, D. T. & Macauley, B. J., 1980. Temporal structure of an ant community in semi-arid Australia. Aust. J. Ecol. 5: 121–134.

Crowell, K. L., 1968. Rates of competitive exclusion by the Argentine ant in Bermuda. Ecology 49: 551–555.

Fox, B. J., 1979. An objective method of measuring the vegetation structure of animal habitats. Aust. Wildl. Res. 6: 297–303.

Fox, B. J. & Fox, M. D., 1978. Recolonization of coastal heath by Pseudomys novaehollandiae (Muridae) following sand mining. Aust. J. Ecol. 3: 447–465.

Fox, M. D., 1978. Changes in the ant community of coastal heath following sand mining. Bull. Ecol. Soc. Aust. 8: 9.

Gillison, A. N., 1978. Minimum spanning ordination – a graphic-analytical technique for three-dimensional ordination display. Aust. J. Ecol. 3: 233–238.

Gower, J. C., 1966. Some distance properties of latent root and vector methods used in multivariate analysis. Biometrika 53: 325–338.

Greenslade, P. J. M., 1971. Interspecific competition and frequency changes among ants in Solomon Islands coconut plantations. J. Appl. Ecol. 8: 323–352.

Greenslade, P. J. M., 1973. Sampling ants with pitfall traps: digging-in effects. Insectes Soc. 20: 343–353.

Greenslade, P. J. M., 1976. The meat ant Iridomyrmex purpureus (Hymenoptera: Formicidae) as a dominant member of ant communities. J. Aust. Ent. Soc. 15: 237–240.

Greenslade, P. J. M., 1979. A Guide to Ants of South Australia. South Australian Museum Special Educ. Bull. Series, Adelaide, 44 pp.

Greenslade, P. & Greenslade, P. J. M., 1971. The use of baits and preservatives in pitfall traps. J. Aust. Entomol. Soc. 10: 253–260.

Greenslade, P. J. M. & Greenslade, P., 1977. Some effects of vegetation cover and disturbance on a tropical ant fauna. Insectes Soc. 24: 163–182.

Haskins, C. P. & Haskins, E. F., 1965. Pheidole megacephala and Iridomyrmex humilis in Bermuda – equilibrium or slow replacement? Ecology 46: 736–740.

Hurlbert, S. H., 1971. The non-concept of species diversity – a cortique and alternative parameter. Ecology 52: 577–585.

Lieberburg, I., Kranz, P. M. & Siep, A., 1975. Bermudan ants revisited: the status and interaction of Pheidole megacephala and Iridomyrmex humilis. Ecology 56: 473–478.

Majer, J. D., 1972. The ant mosaic in Ghana cocoa farms. Bull. Entomol. Res. 62: 151–160.

Majer, J. D., 1978. Preliminary survey of the epigaeic invertebrate fauna with particular reference to ants, in areas of different land use at Dwellingup, Western Australia. Forest Ecol. Manage 1: 32–34.

Majer, J. D., Day, J. E., Kabay, E. D. & Perriman, W. S. (in press). Recolonization by ants in bauxite mines rehabilitated by a number of different methods.

Pontin, A. J., 1961. Population stabilization and competition between the ants Lasius flavus (F.) and L. niger (L.). J. Anim. Ecol. 30: 47–54.

Pontin, A. J., 1963. Further considerations of competition and the ecology of the ants Lasius flavus (F.) and L. niger (L.). J. Anim. Ecol. 32: 565–574.

Room, P. M., 1975a. Relative distributions of ant species in cocoa plantations in Papua New Guinea. J. Appl. Ecol. 12: 47–61.

Room, P. M., 1975b. Diversity and organization of the ground foraging ant faunas of forest, grassland and tree crops in Papua New Guinea. Aust. J. Zool. 23: 71–89.

Simpson, E. H., 1949. Measurement of diversity. Nature 163: 688.

Specht, R. L., Roe, E. M. & Boughton, V. H. (eds.), 1974. Conservation of major plant communities in Australia and Papua New Guinea. Aust. J. Bot. Suppl. 7: 667 pp.

Thom, B. G., 1965. Late Quaternary coastal morphology of the Port Stephens-Myall Lakes area, N.S.W. J. & Proc. Roy. Soc. N.S.W. 98: 23–36.

Williams, W. T., 1976. Pattern Analysis in Agricultural Science. C.S.I.R.O., Melbourne.

CHAPTER ELEVEN

Ant–plant interactions: a world review

Ralf C. Buckley

Introduction

A framework for the analysis of ant–plant interactions has been provided by Hocking (1975) and Gilbert (1979), who grouped insect–plant interactions
into 7 main categories according to the relative roles
of the species involved. I shall consider ant–plant
interactions under the 10 main categories listed
below.

A. *Predation by ants on plants*
   1.Seed harvesting
   2.Leaf cutting

B. *Mutualisms*
   3.Extrafloral nectaries
   4.Food bodies and domatia
   5.Ant-epiphytes
   6.Ant-gardens
   7.Seed dispersal
   8.Pollination

C. *Indirect interactions*
   9.Ant–arthropod–plant systems
   10.Soil modification

These categories are not clear-cut. Seed-harvesting
ants may also disperse seed, for example (Andersen
1982; Majer 1982); domatia may be coupled with
EFN's; and the distinction between domatia, ant-epiphytes and ant-gardens is perhaps more geographic
than functional. Many ant–plant interactions also
involve additional species, such as phytophagous
herbivores which attack undefended plants, or homoptera tended on the host plant as an additional
food source for the ants. There are also interactions in
which ant and plant do not interact directly, but only
through other species. I have separated these as indirect ant–arthropod–plant interactions. They include
cases where competition between ants affects the
interactions of each with plants. Ant nests also modify the plant root environment. Vegetation is of
course also a major component of the ant environment, different ants occupying different vegetation
types, but such patterns are generally inseparable
from the effects of climate, soil, fire regime and insect
fauna on ant distributions and populations, and will
not be elaborated here.

One general feature of ants as a particular class of
animals involved in plant–animal interactions is that
nest sites and food resources are the main factors
limiting their populations (Wilson 1971; Brian 1978;
Brown et al. 1979). Ants have few predators: some
are eaten by birds, lizards, anteaters, ant-lions, spid-
ers and other arthropods (Wheeler 1910; Hatch 1933;
Fall 1937; Holldobbler 1970; Wilson 1971; Cody et
al. 1977; Short 1978; Thiollay 1978; Kilham 1979),
but aggressive interspecific and intraspecific interactions between ants are a much more significant cause
of mortality. Plants provide major food resources for
many ant species, and important nest sites for some.
Hence in many cases ant–plant interactions and ant–
ant competition for plant resources limit ant populations, and this in turn may limit the effect of the
interaction on the plants.

Some types of plant–animal interactions are not
represented in the range of ant–plant interactions.
There do not appear to be any cases where ants
parasitize plants. Nor, apparently, are any higher
plants parasitic on ants, or even on their time and
energy: ants defending or feeding plants or transport-

111

ing their zygotes or gametes apparently always benefit from the interaction. Carnivorous plants catch occasional ants, but do not prey on them specifically.

Various aspects of ant–plant interactions have been reviewed within the last few years (e.g. Carroll & Janzen 1973; Hocking 1975; Harborne 1978; Bernstein 1979a; Brown et al. 1979; Huxley 1980). I shall attempt to analyze each class of interaction as follows. 1. What happens, and where? 2. What plants are involved, and what ants? 3. What are the costs and benefits to each? 4. How species-specific, and how obligate, is the relation? 5. What specializations are involved, and how might the interaction have evolved? These aspects are known in different detail for different interactions, and this is reflected in the depth or brevity of my coverage.

Seed harvesting

Seed harvesting is the prime example of predation on plants by ants. Ants forage for seed on the ground or less commonly on the plant, and take them to the nest to be eaten, stored, or fed to the larvae. In some cases the larvae return nutrients to the workers by trophallaxis (Went et al. 1972). In the case of *Chelaner whitei* and *C. rothsteini* in semi-arid Australia, all the harvested seed are first fed to the larvae, and later returned to the workers as oral or anal secretions (Davison 1982). A wide range of ant species harvest a wide range of seeds in a wide range of habitats (McCook 1879; Tevis 1958; Nickle & Neal 1972; Went et al. 1972; Carroll & Janzen 1973; Levieux & Diomande 1978a, b; Brown et al. 1979; Ashton 1979). Grass seeds are often preferred, perhaps because they are relatively low in toxins (Janzen 1971; Carroll & Janzen 1973; Campbell 1982; Davison 1982), or easy to hull (Campbell 1982), but a wide range of other species are also taken (O'Dowd & Hay 1980; Andersen 1982; Briese 1982; Majer 1982).

The interaction is generally one-sided, though in some cases plants may benefit if seeds are abandoned in shallow underground granaries, where they are protected from fire or other predators. Plant seeds have high food value, being rich in lipids and to a lesser extent proteins (Janzen 1971; Baker 1972; Levin 1973). This value is offset by the costs of finding and harvesting seeds and of overcoming physical and chemical barriers to their consumption, but remains high for consumers with appropriate behavioural,

morphological and physiological specializations. In addition, many temperate and arid-zone plants produce dry seed which can be stored without deterioration in underground granaries, providing a constant food supply despite fluctuations in production. Granivory appears to be particularly common in deserts, despite the great spatial and temporal variations in seed availability (Reichman 1974, 1976; Brown et al. 1979; Davison 1982) and has been adopted by a variety of consumers, notably rodents and birds, as well as ants (Brown & Davidson 1977; Whitford 1979a, b, c; Brown et al. 1979; Morton 1982). Besides seasonal patterns in seed production, there is high variability between years in many arid areas (Went & Westegaard 1949; Tevis 1958; Brown et al. 1979), and annual or ephemeral plants contribute over 80% of the total seed biomass in some regions (Nelson & Chew 1977). Granivorous ants deal with these fluctuations in their food supply by storing food in granaries or a storage caste, and ant populations are only loosely coupled to plant production (Briese 1982; Davison 1982).

For the ants, seed harvesting is a problem in optimal foraging. Ant foraging strategies are therefore an important aspect of seed harvesting. Environment and ant physiology place overall constraints on ant activity, and within these, foraging is influenced by food availability, colony hunger, physical parameters such as light, temperature and humidity, and competition with other granivores. Various foraging strategies are employed in different circumstances, depending on the species and on seed distribution and density, and various cues and signals are used to initiate and control foraging and aid orientation (Holldobbler 1971, 1974, 1976, 1980; Carroll & Janzen 1973; Holldobbler & Wilson 1976; Wheeler et al. 1981).

The range of seed sizes which ants can harvest depends on worker body size. Larger seed are too heavy to lift, whilst smaller seed contain too small an energetic reward to be worth harvesting (Davidson 1978b). Seed shape and surface texture are also important, since they influence handling efficiency (Brown et al. 1975, 1979; Pulliam & Brand 1975). Some ants take inedible floral remnants back to the nest, attached to the seed, either as 'handles' for small smooth seed (Whitford 1978), or perhaps to reduce the time outside the nest, when workers are subject to dehydration and may be attacked by predators or aggressive competitors.

Within the overal range of seeds taken by ants, seeds of different sizes are sometimes taken preferentially by correspondingly sized species, or worker morphs in polymorphic species, and seeds of different shape and surface sculpture by ants with appropriate mandible structure. Morphological differences between *Pogonomyrmex desertorum*, *P. maricopa* and *P. rugosus* in the Upper Sonoran, for example, enable foragers to handle and collect seed of different size and shape (Hansen 1978). Forage and body size are strongly correlated for the six seed harvesters *Pogonomyrmex imberbiculus*, *P. desertorum*, *P. rugulosus*, *Pheidole rugulosa*, *Ph. xerophila*, and *Ph. militicida* in southeastern Arizona (Chew & de Vita 1980), and all members of the ant granivore guild differ in size, with a mean body weight ratio of 1.66 between each species and its nearest neighbour (Chew 1977). Davison (1982) found that though *Chelaner whitei* and *C. rothsteini* in semi-arid Australia collected seeds in the same overall size range, the smaller *C. rothsteinii* collected proportionately more of the smaller seed. In addition, each species is most active when its preferred seeds are most abundant. Similarly Whitford (1978a) found that there is little overlap in worker body length between the members of a local foraging guild, and where species overlap in body size they differ in diel and seasonal activity patterns. Rissing (1981), however, found that worker size accounts for <4% of the variance in the size of seeds chosen by *Pogonomyrmex rugosus* and *Veromessor pergandei* at a site in the Nevadan Mojave Desert. If presented with a mixture of seed species, colonies, of each ant species took some of each, but individual ants specialized on particular seed species. Again, Morton (1982) found that several polymorphic genera could coexist in sites in the Australian arid zone.

These patterns reflect competition within foraging guilds, and Davidson (1977) suggested that desert ant granivore communities are 'structured on the basis of competition'. In general the available evidence supports this, at least for the Sonoran Desert. Hansen's (1978) study of *Pogonomyrmex desertorum*, *P. maricopa* and *P. rugulosus* in the Upper Sonoran, for example, showed that the three species differ in worker physiology, particularly desiccation resistance, which enables them to forage at different times and to consume different proportions of non-seed food, mainly arthropods. They also nest in different microhabitats, giving access to different food as-semblages. *P. rugosus* and *P.barbatus* take similar forage, but whilst *P.barbatus* is most active in early summer and forages at night, *P.rugosus* is more active in mid summer and forages at night only in July (Whitford et al. 1976). Where *P.rugulosus*, *P.desertorum* and *P.californicus* co-exist, seed selection by the first two is related to foraging strategy, but *P.californicus* takes seed so as to avoid the other species (Whitford 1978). Minor competitors of the dominant *P.desertorum*, however, may do better where it is under strong competition from *P.rugosus* (Davidson 1977a). *Veromessor pergandei* produces more morphs where it has fewer competitors (Davidson 1978a), enabling it to take a wider range of seed.

The above evidence derives largely from the North American deserts, but similar patterns have been found elsewhere. On the Ivory Coast, for example, *Messor galla* forages at night in December and takes seed from the short and medium grasses *Monocymbium seresiiforme* and *Pennisetum hordeoides*, whereas *M. regalia* is diurnal and concentrates on the tall *Andropogon gayanum* (Levieux & Diomande 1978a). In *Eucalyptus regnans* forest with *Pomaderris aspera* understorey in central Victoria, Australia, foraging activity of *Iridomyrmex biconvexus*, *Prolasius pallidus* and *P.brunneus* is separated diurnally, seasonally and spatially (Ashton 1979). Similarly Lynch et al. (1980) observed foraging segregation between *Aphaenogaster rudis*, *Paratrechina melanderi* and *Prenolepis imparis* occurring sympatrically in a hardwood forest in Maryland, U.S.A.

Competition between colonies of individual species is also common amongst seed-harvesting desert ants. Bernstein & Gobbel (1979) suggest that intraspecific territory defence, intercolony aggression and colony overdispersion are characteristic of desert ants, whilst interspecific territory defence is common in more tropical ant communities, which contain more species but fewer individuals of each. Internidal aggression is usual in *Veromessor pergandei* (Wheeler & Rissing 1975). In the Australian arid zone some groups of *Calomyrmex* nests are compatible whilst others, presumably from different colonies, are not (Brough 1976). Nests are regularly dispersed in *Veromessor pergandei*, *Pogonomyrmex rugosus* and *P.californicus*, indicating intraspecific colony defence (Bernstein 1975; Holldobbler 1974, 1976). Whitford et al. (1976), however, found that though there was no overlap between the colonies or foraging territories of *P.rugosus* and *P.barbatus*, and strong aggression between the two species, there was no aggression between the colonies of each species, and considerable overlap in foraging territories.

Seed harvesting by ants is also modified by competition with other granivores, particularly rodents (Brown & Davidson 1977; Brown et al. 1979). Rodents take more seed than ants in the North American deserts (Brown et al. 1975, 1979; Mares & Rosenzweig 1978), consuming up to 90% of the total *Larrea* seed crop (Chew & Chew 1970), 30-80% of seed in experimental plots (Nelson & Chew 1977), and 95% of the *Erodium* seed crop (Soholt 1973). Sympatric ants and rodents subdivide the overall seed resources (Davidson 1977a, b; Reichman 1979), largely by differential foraging patterns in relation to different seed densities and distributions (Reichman & Oberstein 1977; Price 1978a, b; Reichman 1979). This diversity in foraging technique contributes to the high diversity of ants and rodents in deserts (Brown et al. 1975; Brown & Davidson 1977; Davidson 1977a, b). Rodents are intensive granivores (Reichman 1975, 1977), and are more efficient than ants at finding and harvesting large seed clumps (Brown et al. 1975, 1979). Unlike ants, they can also forage beneath the soil surface (Reichman 1979), and besides competing for the same seed initially, heteromyids such as *Dipodomys* in Nevada compete further with *Pogonomyrmex occidentalis* by digging up their seed stores in spring (Clark & Comanor 1973a).

Ant foraging activity is strongly influenced by physical environmental conditions, particularly light, temperature and humidity (Sudd 1967; Whitford & Ettershank 1975; Davidson 1977a; Whitford 1978; Bernstein 1979b). Some ants are nocturnal foragers, others diurnal, and some, such as *Pogonomyrmex rugosus,* forage diurnally in winter and nocturnally in summer (Whitford & Ettershank 1975; Whitford et al. 1976). Others forage in the morning and afternoon but not at noon or night: for example, *Veromessor pergandei* (Clark & Comanor 1973) and *Brachyponera senaarensis* (Levieux & Diomande 1978b). Different species forage within different temperature ranges (Tevis 1958; Schumacher & Whitford 1974; Whitford & Ettershank 1975; Brown et al. 1975, 1979; Bernstein 1975; Johns & Greenup 1976; Mott & McKeon 1977); minimum foraging temperatures range from 1.7 °C for *Prenolepis imparis* to 23.8 °C for *Messor aegyptiacus,* and maxima from 23.9 °C for *P.imparis* to over 60 °C for Austral-

ian *Melophorus* species (Talbot 1943; Ayre 1957, 1958; Sheata & Kaschef 1971; Bernstein 1979b; Morton 1982). Ant species also differ in their tolerance to desiccation: in the Chihuahuan Desert, for example, foraging by *Formica perpilosa* is largely independent of saturation deficit, whilst *Trachymyrmex smithi neomexicanus* forages only when the saturation deficit is below 35 gm.m (Schumacher & Whitford 1974). *F.perpilosa* feeds on water-rich plant exudates and so can afford to forage during the most arid conditions, although the desiccation rate is higher and desiccation tolerance lower for its workers than for those of sympatric seed-eating species of *Novomessor* and *Pogonomyrmex* (Whitford et al. 1975).

Foraging activity is also controlled by food availability. In New Mexico, for example, the seasonal foraging patterns of granivorous ants vary from year to year, depending on plant production and hence on rainfall (Whitford 1978a). Again, where *Veromessor pergandei, Pogonomyrmex rugosus* and *P.californicus* are distributed differentially along an altitudinal gradient, eacht species forages most actively at the time of peak seed production at its own preferred altitude (Bernstein 1975). In seed-storing species such as *Pogonomyrmex rugosus,* foraging ceases when the granaries are full (Whitford & Ettershank 1975). In semi-arid western New South Wales, Australia, *Chelaner whitei* uses its final-stage larvae as a storage caste rather than storing seed in granaries, and foraging activity is inversely related to larval biomass (Davison 1982).

Different species employ different foraging techniques, and individual species may use different methods at different food densities. Techniques may be divided broadly into individual and group foraging, and the latter subdivided according to recruitment and orientation cues, etc. Column foraging removes clumped seed efficiently, whilst individual foraging is more efficient for dispersed seed (Davidson 1977a), whether the vegetation is open or dense (Holldobbler 1976; Whitford 1976). *Veromessor pergandei* forages in long narrow columns when food density is low, but individually when food density is high (Bernstein 1975a). The foraging columns of *V.pergandei* in Death Valley maintained constant directions in 1973 but rotated in 1974, in response to changes in food density (Rissing & Wheeler 1976); Whitford (1976) found that some colonies of *Pogonomyrmex rugosus* employed group foraging, others individual. In the Australian arid zone species of *Chelaner, Mono-*

morium and *Pheidole* forage in columns, whilst *Melophorus, Meranoplus, Tetramorium* and *Rhytidoponera* forage individually (Morton 1982).

The impact of granivorous ants – the cost to the plants – can be considerable, and in some areas they represent an agricultural problem of considerable economic significance (Campbell 1982). Colony size and density are extremely variable, but Whitford & Ettershank (1975), for example, gave an estimate of 180,000 *Pogonomyrmex* foragers per hectare in a desert area, whilst Ashton (1979) estimated that there were 5 or 6 million seed-harvesting ants per hectare in Australian *Eucalyptus regnans* forest. Granivores are often a major control on seed density (Janzen 1960, 1971; Harper 1978), and ants and rodents are the principal seed consumers in the North American desert (Brown et al. 1975, Whitford & Ettershank 1975; Davidson 1976; Brown & Davidson 1977; Brown et al. 1979a, b). Reichman (1979) showed that either ants, rodents, or both could decrease soil seed load substantially. Seed harvesting is generaly highly selective, however; the overall proportion of seeds removed is relatively small, but particular plant species may lose a large proportion of their seed crop to ants (Tevis 1958; Box 1960; Willard & Crowell 1965; Eddy 1970; Went et al. 1972; Clark & Comanor 1973; Rogers 1974; Whitford 1975, 1978; Withers 1978). Thus Tevis (1958) estimated that *Veromessor pergandei* took 37 million seeds per hectare per year in the Mojave Desert, amounting to 1% of the total seed crop, and Rogers (1974) estimated that *Pogonomyrmex rugosus* takes 2% of available seed biomass in a shortgrass prairie in north-eastern Colorado. *P.occidentalis* cuts 157–226 million annual plants per hectare per year in Nevada (Clark & Comanor 1973), in *Eucalyptus marginata* forest in Western Australia, approximately 2.7% of legume seeds are in ant nests (Majer 1982). Near Katherine in the Northern Territory of Australia, seed removal rates by *Pheidole* species have been estimated (Mott & Mc Keon 1977) as approaching the 7,000 seeds per colony per day quoted by Tevis (1958). In western New South Wales, however, soil seed reserves of 3 common grasses did not decrease detectably during 16 months of drought, despite continued harvester-ant activity, perhaps since seeds are buried faster in the cracking clay loams of the Australian study site that in the sandy and gravelly soils of North American sites (Westoby et al. 1982a). The impact of seed harvesters in the arid zone depends on past years'

rainfall (Whitford 1978), since this influences seed production and colony hunger: *Pogonomyrmex* species in the Chihuahuan Desert took 10% of seed production in the first wet year after a drought, but less than 3% in the following year when granaries had been replenished (Whitford 1975, 1978b, c).

Selection of seed for energy efficiency is likely to be an optimal strategy only at times of extreme colony hunger (Brown et al. 1975; Pulliam & Brand 1975); at other times it may be preferable to select seed to maximize intake of a specific nutrient (Pulliam 1974) or to minimize intake of toxins (Whitford 1978b). About 15% of annual and herbaceous perennial dicots contain tannins (Rhoades & Cates 1976). Seed selectivity is expected to decrease when food is scarce (MacArthur & Pianka 1966; Schoener 1971), and this was confirmed for *Pogonomyrmex rugosus* by Whitford (1978b). Forage selection in this species is not controlled only by energy efficiency or by seed size or abundance (Davidson 1977a). *Pogonomyrmex* and *Pheidole* generally prefer annual grasses and forbs (Willard & Crowell 1965; Clark & Comanor 1973; Rogers 1974; Whitford 1975, 1978; Mott & McKeon 1977; Ballard & Preuss 1979); *Pheidole* species in the Katherine area take the native Australian grasses *Themeda australis* and *Digitaria ciliaris* in preference to the introduced legume *Stylosanthes humilis* (Mott & McKeon 1977). Foraging preferences may change seasonally: *Pogonomyrmex* species in North America, for example, take a variety of annuals early in the season but switch to *Bouteloua barbata* as soon as it seeds (Whitford 1978b). Both *Chelaner whitei* and *C.rothsteini* in semi-arid Australia also show seasonal changes in forage selection, taking greater proportion of ephemeral forb and grass seeds when these are abundant (Davison 1982). Whitford (1978a, b, c) contrasted the seed selection strategies of three *Pogonomyrmex* species. *P.rugosus,* a group forager, preferred plants releasing large numbers of seed simultaneously; *P.desertorum,* foraging individually, took the most abundant seed at any given time, but with a preference for grass seed later in the growing season; and *P.californicus* selected seed which enabled it to avoid the other two species. Similarly, *Chelaner rothsteini* takes more grass seed than *C.whitei* (Davison 1982), whilst *C.whitei* takes more composite and chenopod seed.

Nickle & Neal (1972) suggested that selective foraging by *Pogonomyrmex badius* was 'apparently in response to chemicals in the seed', and Ashton (1979) that *Eucalyptus regnans* seed in central Victoria contain sweet substances attractive to ants. Withers (1978) also found that seed-harvesting ants took large numbers of *Eucalyptus ovata* and *E.leucoxylon* seed, ignoring these of the understorey *Casuarina littoralis*. Since *Prolasius banneus*, *P.flavicornis*, *P.pallidus* and *Chelaner leae* together took over 60% of *E.regnans* seed production in normal years, and perhaps the entire seed crop, Ashton (1979) suggested that the differential regeneration of *E.regnans* after fire was not due any effects of the fire on seed germination but to its interference with seed-harvesting ants and to their saturation by a massive fire-stimulated seed fall. A similar mechanism operates for *E.incrassata* (Wellington 1981).

How dependent on seed are seed-harvesting ants? Many ants take seed as part of an omnivorous diet (Creighton 1953; Went et al. 1972; Whitford & Ettershank 1975; Whitford et al. 1976; Whitford 1978a, b; Brown et al. 1979) and are not dependent on them. The seed-harvesting *Brachyponera senaarensis* on the Ivory Coast can take up to 25% arthropods (Levieux & Diomande 1978b). Even strict desert granivores such as *Pogonomyrmex*, *Novomessor* and *Pheidole* will occasionally prey on *Amitermes* and *Gnathamitermes* when these are foraging at the surface (Whitford 1978b). It seems that most granivorous ants are dependent on seed, at least to maintain present population levels, but that few, if any, can survive on seed alone (Brown et al. 1979). In addition, many species now have morphological, physiological and behavioural adaptations for granivory. Hence overall, many granivorous ants are now largely dependent on seed. Such dependence presumably evolved gradually, however: collecting seeds as part of an omnivorous diet does not necessarily require morphological or behavioural specializations, and species which now possess such adaptations need not have crossed any major evolutionary barriers in acquiring them.

Leaf cutting

Some tropical ants feed on fungi which they cultivate on macerated plant material in special nest chambers. These ants cut sections of leaves, and to a lesser extent flowers and fruits, from a variety of plant species. They transport them to the nest chambers, where they defaecate on them and inoculate them with fungus. When the fungus has exhausted available nutrients the ants discard the chamber contents into water or refuse dumps. Hence the ants gain food, indirectly, from the cut material. The plants lose primary production and photosynthetic material and are rendered more susceptible to secondary predators and pathogens. Owen (1980) has argued that some herbivores may benefit plants by accelerating nutrient cycling, but leaf cutting is unlikely to benefit the cut plants since cut material is taken beyond their root zone. Refuse dumps modify soil nutrient status (Haines 1975), which may benefit some seedlings, but generally not those of the cut species. Hence the interaction is one-sided.

The main leaf-cutting ants are species of *Atta* and *Acromyrmex*. The overall range of the Attini stretches from 4° N to 44° S (Weber 1972a), but leaf cutting appears to be confined to the tropics. There is a broad division into forest-dwelling leaf cutters and savanna-dwelling grass cutters. Other genera involved include *Cyphomyrmex*, *Mycetophylax*, *Myrmicocrypta* and *Trachymyrmex* (Weber 1968). Leaf-cutting ants utilize a wide variety of plant species.

The foraging territories of tropical forest attines are comparatively large (Cherrett 1968; Haines 1975; Hubbell & Rockwood 1980), the largest reported radius being 250 m for *Atta cephalotes* (Lewis et al. 1974a). Diel patterns of foraging activity vary according to season for *Atta colombica* and *A. cephalotes* in Costa Rica (Rockwood 1974) and *Atta sexdens* in subtropical Paraguay (Fowler & Robinson 1979), though Lewis et al. (1974b) found that diel patterns in the foraging of *Atta cephalotes* were related to brood size rather than physical environmental factors. The intensity of foraging activity also varies (Fowler & Robinson 1979): maximum foraging intensity by *Atta colombica* and *A.cephalotes* in Costa Rica, for example, coincides with peak production of new leaves and flowers (Rockwood 1975). Attines maintain well-defined foraging trails, marked with pheromones (Moser & Blum 1963; Wilson 1971), which may be maintained by unladen workers (Lugo et al. 1973; Fowler 1978). Trail lengths vary seasonally for *Atta sexdens* (Fowler & Robinson 1979).

Foraging labour is divided in some attines. In subtropical Paraguay, for example, one group of *Atta sexdens* workers cuts leaf fragments from the canopy; a second recovers them, cuts them up, and carries them to the main foraging trail; and a third takes them back to the nest (Fowler & Robinson 1979). *Atta cephalotes* also employs a 'bucket-brigade' system where foragers transfer leaf fragments to carrier ants at the junctions of new branch trails and larger established trails (Hubbell et al. 1980), whilst the 'cut-and-collect' system was first noted by Daguerre (1945) for *Atta saltensis*. Such a multistage foraging method will be energetically preferable to carrying fragments down the trunk if the costs of lost fragments are less than the costs of return journeys up the trunk (Hubbell et al. 1980). Fowler & Robinson (1979) found a recovery efficiency of 49% for the *Atta sexdens* three-stage system, and pointed out that its impact on the vegetation would therefore be twice that estimated from the loads entering the nest.

The desert attines also maintain large colonies, those of *Atta mexicana* being the largest ant colonies in the North American deserts (Mintzer 1979). This species constructs a network of underground foraging tunnels, and its foraging territories may cover 8,000 m² (Mintzer 1979). It forages by day in winter and at night in summer. Workers of the desert gardening ant, *Acromyrmex versicolor versicolor,* forage individually as well as along the typical attine trails (Gamboa 1975). Mintzer (1980) found adjacent colonies of *Acromyrmex versicolor* and *Atta texana* using the same foraging trails with no evidence of aggression except at the *A.versicolor* colony entrance, despite a strong overlap in forage preferences. Jutsum et al. (1979) found intercolony aggression in *Acromyrmex octospinosus,* increasing with colony separation, and showed by means of laboratory experiments that the differences in colony odour which triggered aggression were due more to diet than to genetic differences.

The quantitative impact of leaf-cutting ants can be considerable. Grazing damage to Paraguayan pastures is correlated with colony density of *Acromyrmex landolti* (Fowler & Robinson 1977). Lugo et al. (1973) calculated that leafcutting by *Atta colombica* in Panama reduced forest primary productivity by 1.76 kcal/m² day. The energy flow through their nests was 1.7% of the total annual energy flow through leaf litter fall, per unit forest area (Haines 1978), and corresponding proportions for 13 nutrients ranged from 0.5 to 3.1%. This material is ultimately channelled through the refuse dumps, increasing their nutrient status (Lugo et al. 1973; Haines 1975); in this case, annual nutrient flows were 16–98 times the flows in leaf litter when expressed per unit area of

dump (Haines 1978). Root density was increased four-fold within the dumps, and was particularly high within the top 20 cm (Haines 1978). The 0–20 cm layer of *Atta colombica* dump soils on Barro Colorado Island contained 3 500 ppm, total N, as compared to 2 800 ppm in shallow forest floor soils; 17 cf. 0.2 ppm available P, 700 cf. 300 ppm total P, 1 200 cf. 100 ppm total K, 2 000 cf. 1 000 ppm total Mg, and 6 000 cf. 2 000 ppm total Ca (Haines 1975). Test plants grew 2–6 times faster on shallow dump soils than on shallow soils from the nests or the forest floor. Forest succession is probably also modified by *Atta* (Haines 1975). There is also a modified or accelerated succession on abandoned attine nests; those of *Atta vollenweideri* can act as nuclei for woodland in Paraguayan chaco pastures, for example, being invaded by three *Prosopis* species (Jonkman 1978, 1979).

Attine leaf harvesters forage selectively, and the impact on particular species can be much greater than on the plant community as a whole. This is of particular economic relevance since introduced crops such as citrus are often attacked selectively. In native forest in Guyana, *Atta cephalotes* takes material from 36 of the 72 available plant species, concentrating on a few of these (Cherrett 1968). In Costa Rica, *A.colombica* and *A.cephalotes* again sample many species, but take mature leaves only from 31% and 22% of available species respectively, together with new leaves from 12–16 further species (Rockwood 1976). Of these, over 80% of the leaves are taken from only 10 and 6 plant species respectively (Rockwood 1972, 1976; Rockwood & Glander 1979). Laticiferous plants are generally avoided. In Paraguayan pastures, *Acromyrmex landolti fracticornis* preferentially selects *Digitaria smutsii* and *Panicum maximum* from 13 introduced forage grasses (Fowler & Robinson 1977); and in the Sonoran Desert, *Atta mexicana* concentrates heavily on two winter annuals when these are present (Gamboa 1975), and on a small group of perennials otherwise.

Attines forage differentially for different plant parts. *Atta colombica* and *A.cephalotes* in Costa Rica selectively harvest new leaves and flower parts, in addition to the mature leaves of particular species and some fruit fragments (Rockwell 1975; Rockwell & Glander 1979). In Paraguay, *Acromyrmex landolti* selectively harvests the more tender grass blades (Fowler & Robinson 1977). There are seasonal differences in selectivity: *Atta colombica* in Costa Rica takes more new leaves in the dry season and more floral material in the wet (Rockwood 1975), and on a more opportunistic basis, *Acromyrmex versicolor* harvests more fresh plant material after wet periods and more dry grass after droughts (Gamboa 1975).

Attines generally select plant tissues with high moisture content (Cherrett & Seaforth 1968; Barrer & Cherrett 1972; Haines 1971, 1975; Rockwood 1972). Recently Bowers & Porter (1981) studied forage selection by *Atta columbica* on Barro, Colorado Island in detail, and showed firstly that tissue moisture content and palatability are often related, and secondly that ant selection of plant tissues with a given moisture content depends on tissue palatability and distance from the nest.

Leaf-cutting ants generally do not attack the closest individuals of preferred species as intensively as more distant plants, and may not attack the closer plants at all. Cherrett (1968), for example, found that a colony of *Atta cephalotes* in Guyana left, on average, 6.2 intact specimens of any preferred species between the nest and the first specimen attacked. On the basis of exploitation patterns he suggested that this species employed a 'conservative grazing system'. Subsequently, however, Fowler & Stiles (1980) concluded that 'apparent conservative management of resources by leaf-cutting ants is more easily explained by the response of foraging workers to the patchiness of the vegetation being harvested,' and that 'foraging trails are built and maintained to permit an ordered area-restricted search for palatable vegetation'.

Foraging preferences appear to be chemically mediated (Littledyke & Cherrett 1975). *Atta cephalotes* and *Acromyrmex octospinosus* tested in a moving-air olfactometer responded positively or negatively to the odours of various substrate materials (Littledyke & Cherrett 1978b), and some non-attractive leaves were rendered attractive by removing the cuticular wax. Pith segments impregnated with extracts of young leaves are preferred to those impregnated with extracts of old leaves, and non-lipid extracts preferred to extracts containing lipids (Littledyke & Cherrett 1978a). Selectivity decreases during periods of high foraging activity (Littledyke & Cherrett 1975).

Though attine workers consume sap from cut leaf edges (Stradling 1978), leaves are cut primarily for the fungus gardens. Hence leaf selection depends on the physiology of the fungus as well as that of the ants. The food fungus of *Atta cephalotes,* for example, grows faster on plant extracts preferred by the ant than on other extracts (Mudd & Bateman 1979). The taxonomy and physiology of the fungal symbionts have been described by Hervey et al. (1977). The fungi provide the ants with external cellulases, but require careful gardening to be retained in monoculture. All attines defaecate on the substrate before adding it to the garden (Weber 1958), and all attines so far tested contain chitinases in their faecal material (Martin et al. 1973); this prevents invasion of the gardens by fungi containing structural chitin. The garden fungi themselves may produce antibiotic metabolites, as for example the *Lepiota* species cultivated by *Cyphomyrmex costatus* (Hervey & Nair 1979; Nair & Hervey 1979). Attine gardens contain a variety of extraneous insects (Waller 1980), but ants control any fungivores (Quinlan & Cherrett 1978a). Garden temperatures are also regulated; the fungus cultured by *Acromyrmex octospinosus,* for example, growing best around 25 °C (Quinlan & Cherrett 1978).

The fungi produce clusters of swollen hyphae known as staphylae, which are eaten by workers and fed to the larvae. The ants prefer staphylae to ordinary hyphae (Quinlan & Cherrett 1978) and their own fungi to others. The fungal cell contents would comprise a balanced diet for the ants (Martin et al. 1969; Weber 1972), and it was once thought that the fungus was the ants' sole food (Belt 1874; Weber 1972). Experiments using radio-labelled phosphorus, however, have shown that workers of *Atta cephalotes* take sap from the edges of cut leaves (Stradling 1978). In addition, measured consumption rates indicate that the workers' respiratory energy requirements would not be satisfied by eating staphylae alone (Quinlan & Cherrett 1979), though those of the larvae would. In now appears that workers satisfy their energy requirements by imbibing plant sap whilst cutting leaves and chewing them for the gardens, as well as by eating staphylae (Quinlan & Cherrett 1979). Workers cannot consume solids owing to the filtering action of the infrabuccal pouch (Quinlan & Cherrett 1978b). The fungus, however, is the sole larval food and is essential to colony maintenance.

Leaf-cutting ants are thus totally dependent on their fungus, but not on any one food plant to supply the gardens. This is also of considerable economic importance: leaf-cutting ants attack a variety of crops, notably citrus (Mudd et al. 1978), but can be controlled by placing leaves of the jackbean, *Canavalia ensiformis,* on their nests: the ants incorporate these leaves in the gardens, and it appears that the fungicide demethyl homopterocarpin, present in the leaves, kills the fungus and hence the nest (Mullenax 1979). A similar control technique involves incorporation of microencapsulated dioxathion or permethrin in citrus pulp, which the ants take into the nest (Samways 1981). An alternative control technique has been suggested in Brazil, where *Atta sexdens* defoliates *Eucalyptus* forests: a scarabeid beetle, which consume young *Atta* queens as they burrow into the ground, is being tested as a biological control agent (Samways 1981).

What defence do the plants have against leaf-cutting ants? Since attines are generalist herbivores, quantitative rather than qualitative defences would be expected (Freeland & Janzen 1974; Feeny 1975, 1976; Cates & Orians 1976; Rhoades & Cates 1976). Cuticular waxes (Littledyke & Cherrett 1978b) and latexes (Stradling 1978) appear to be effective deterrents; so too would be fungicidal compounds. The selection of young leaves and floral parts rather than old leaves also supports this, since quantitative defences are generally more strongly developed in mature leaves. There is also intraspecific variation in chemical defences (Glander 1975a, 1977; Gates 1975; Mothes 1976).

Attine gardens involve a tight and specialized interaction between the ants and their fungi, but a relatively unspecialized and one-sided interaction between the ants and the higher plants whose leaves they cut. The evolution of attine leaf cutting and fungus gardening has been discussed by Wheeler (1907), Weber (1958), Wilson (1971) and Garling (1979), who suggests that ant gardens originated from the consumption of mycorrhizal hyphae on roots in underground chambers.

Extrafloral nectaries

A wide range of plants in both temperate and tropical latitudes produce extrafloral nectar which attracts ants. The ants collect and consume the nectar and defend the nectaries, or the whole plant, against phytophagous insects and mammalian herbivores, reducing damage to the plants (Janzen 1966a, b, 1967; Hocking 1970; Elias & Gelband 1975; Inouye & Tay-

lor 1975, 1979, 1980; Keeler 1975, 1977a, b, 1980a, b; Bentley 1976, 1977a, b; Deuth 1977; Tilman 1978; Schemske 1978, 1980; Koptur 1979; O'Dowd 1979; Pickett & Clark 1979; Blom & Clark 1980; Elias 1980; Scott 1981). The subject has been reviewed recently by Bentley (1977a), so the present summary does not consider early literature in any detail.

Though the earliest descriptions of interactions between ants and plants possessing extrafloral nectaries assumed that the ants performed a protective function (Delpino 1874; von Wettstein 1889), this 'protectionist' hypothesis was questioned by Wheeler (1910), and only three years ago Vogel (1978) reiterated the alternative that 'extra-floral nectaries are primarily excretory organs.' A historical account of the debate was given by Gottsberger (1972) and summarized by Bentley (1977a) and Inouye & Taylor (1979). The evidence against the 'excretory' hypothesis now seems overwhelming.

The effectiveness of an ant defence in reducing herbivore damage was first shown by Janzen (1967) for swollen-thorn acacias and there have been many subsequent demonstrations for EFN plants: e.g. Bentley (1977b) for *Bixa orellana*, Pickett & Clark (1979) for *Opuntia acanthocarpa*, Koptur (1979) for *Vicia sativa* and *V. angustifolia*, Inouye & Taylor (1979) for *Helianthella quinquenervis*, and Keeler (1980a) for *Ipomoea leptophylla*. Ant defence of *H. quinquenervis* is more effective at higher altitudes (Inouye & Taylor 1979); at lower altitudes, both seed predators and ants are more abundant and the ants are less effective.

The extent of ant activity ranges from local defence of the nectaries to regular patrols over the entire plant. Ant activity may vary with the time of day or with soil or plant temperature, as shown by Blom & Clark (1980) for *Ferocactus gracilis* in Mexico. Typically the ants show 'ownership behaviour' (Way 1963), actively attacking or pursuing intruding insects (Wheeler 1910; Ayre 1968; Wilson 1971; Inouye & Taylor 1979). The mere presence of ants on a plant often reduces the abundance of other insects (Gotwald 1968). Arboreal ants in general will often move their nests closer to sugar sources (Bentley 1977a), as in the case of *Oecophylla* (Brian 1955), and the attendant ants nest permanently in or on some EFN plants: in the pithy stems of *Bixa orellana* (Carroll 1974; Bentley 1977b), or the hollow internodes of *Leonardoxa africana* (Elias 1980), for example. Cavities in *Acacia saligna* are inhabited by

Crematogaster species (Morellini 1977; Majer 1979), and another *Crematogaster* species builds shelters around the fruits and ovaries of *Ipomoea pandurata* (Beckman & Stucky 1981).

The ant defence may reduce damage to leaves (Bentley 1976a, b; Keeler 1977, 1980; Tilman 1980; Koptur 1979; O'Dowd 1979); to flowers (Elias & Gelband 1975; Keeler 1977, 1980; Beckman & Stucky 1981; Scott 1981); or to seed and fruit (Bentley 1976; Schemske 1978, 1980; Inouye & Taylor 1979; Pickett & Clark 1979; Keeler 1980a). The ants may also reduce flower robbing (Elias & Gelband 1975): nectaries on the sepals of *Ipomoea carnea,* for instance, attract ants which reduce nectar robbing by large *Xylocopa* bees (Van der Pijl 1954; Keeler 1977). Alternatively, they may reduce corolla damage, as on *Ipomoea leptophylla* (Keeler 1980a) and *I.pandurata* (Beckman & Stucky 1981). In the absence of ants, grasshoppers generally damage 100% of *I.leptophylla* corollas by midday (Keeler 1980b). Insects also damage the buds of *Bixa orellana,* and fruit set is greater for plants with more ants (Bentley 1977b). *Helianthella quinquenervis* are attacked by the larvae of tephritid flies, an agromyzid fly and three Lepidoptera; ants successfully interrupt oviposition by flies but are ineffective against Lepidoptera or mammalian herbivores (Inouye & Taylor 1979). Fruit set is lowered in plants of *Vicia sativa* and *V.augustifolia* from which the extrafloral nectaries have been excised (Koptur 1979), and ants visiting the sepal nectaries of *Ipomoea leptophylla* decrease seed loss to bruchid beetles, particularly *Megacerus discoidus* (Keeler 1980a).

At least 20 ant genera have been reported in interactions with EFN plants (Schemske 1978; Tilman 1978; Inouye & Taylor 1979; Keeler 1979a; Koptur 1979; O'Dowd 1979; Pickett & Clark 1979; Blom & Clark 1980; Elias 1980), and extrafloral nectaries also attract wasps, flies and beetles (Putnam 1963; Keeler 1978). The ants defend their 'host' plants against a wide variety of insect herbivores; the most important seem to be flies, such as *Euxesta* sp. on *Costus woodsonii* (Schemske 1980) and those mentioned above for *Helianthella quiquenervis;* Lepidoptera, such as *Malacosoma americanum* on *Prunus serotinus* (Tilman 1978) and others on *Ipomoea leptophylla* (Keeler 1980a); and bruchid beetles: others such as grasshoppers have already been mentioned (Keeler 1980a). Majer (1979) collected 112 non-ant insect species from trees of *Acacia saligna* using sweep samples, and of these 69 are phytophagous.

'A great diversity of plant species in a wide variety of habitats' possess extrafloral nectaries (Schemske 1980); according to Keeler (1979b) they are known from 73 angiosperm families and a few ferns, though Elias (1980) gives a figure of only 66 families. Plants possessing extrafloral nectaries are found worldwide (e.g. Zimmerman 1932; Schnell et al. 1963; Bentley 1977a; Keeler 1979; Elias 1979), in both temperate regions (e.g. Tilman 1978; Inouye & Taylor 1979; Majer 1979; Keeler 1979b, 1980a; Koptur 1979) and the tropics (e.g. Janzen 1966a, b; Bentley 1977a, b, 1979; Keeler 1977, 1979a; Schemske 1978; Pickett & Clark 1979; O'Dowd 1979). It has been suggested that interactions involving extrafloral nectaries are more common in the tropics (Gilbert in Orians 1974; Bentley 1977a) but currently available data are insufficient to assess this (Keeler 1979b), and the effects of latitude are confounded with those of altitude. The proportion of species with extrafloral nectaries ranges from 0–14% in temperate zone communities so far studied (Keeler 1980a) and from 0–80% in the tropics (Bentley 1979; Keeler 1979a). This proportion is correlated with ant abundance at both latitudes (Bentley 1976, 1977a, 1979; Keeler 1979a, 1980a, b). The proportion varies with the plant community: in Nebraska, for example, it is 1.3% in riparian forest understorey, 1.8% in virgin deciduous forest understorey, 0.0% in tallgrass prairie, and 8.3% in sandhill prairie; whilst in California EFN's have not been recorded from native plants at all (Keeler 1980b), but only from introduced species (Koptur 1979). Again, the proportion ranges from 0.3 to 0.8 in dry forest in Guanacaste (Bentley 1976) but only 0.1 to 0.4 in riparian forest. Within the tropics, ants are more abundant at lower altitudes (Janzen et al. 1976; Carroll & Van der Meer 1972), though the precise patterns differ between regions. In Jamaica the proportion of plants with extrafloral nectaries falls from 28% at sea level to none at 1,310 m; ant abundance also falls with altitude, but there is still a considerable ant fauna at the higher elevation (Keeler 1979a). Though there is a general overall correlation between ant abundance and the frequency of extrafloral nectaries, therefore, the patterns are not clearcut.

Extrafloral nectaries may be situated on various parts of the plant, and it seems likely that they evolved repeatedly (Elias 1980). In Nebraskan prairie plants they are commonly on the leaves (Keeler 1979b); in *Ipomoea pandurata* they are on the upper part of the pedicel (Beckman & Stucky 1981); in *Helianthella quiquenervis* and *Alyogyne hakeifolia* on the involucral bracts (Inouye & Taylor 1979; Scott 1981); in *Vicia sativa* and *V.angustifolia* on the stipules (Koptur 1979); and in Australian phyllodinous acacias typically on the upper margin of the phyllodes near the distal end of the pulvillus (Majer 1979; Buckley ms). *Ochroma pyramidale* possesses two types of foliar nectary, on leaf-vein and petiole respectively (O'Dowd 1979), and *Campis × tagliabuana* possesses 4, situated respectively on petioles, sepals, petals and developing fruits (Baker et al. 1978). In general EFN's are external or partly embedded (e.g. Keeler & Kaul 1979), but those of *Leonardoxa africana* are completely embedded in the leaf mesophyll, opening to the abaxial surface through a narrow neck (Elias 1980). They are more common on younger leaves (O'Dowd 1979; Elias 1980); and in *Acacia saligna* at least, they are more active in younger plants (Majer 1979). In *Populus grandidentata* they are more prominent and active on the later of each year's two leaf flushes (Curtis & Lersten 1978).

In addition to strictly extrafloral nectaries, some plant species possess postfloral nectaries: nectaries that first serve to attract pollinators but continue to function subsequently, attracting ants which defend them. Postfloral nectaries have been described by Daumann (1932, 1974), Bentley (1977a) and Faegri & Van der Pijl (1979), but their function was first demonstrated by Keeler (1981) for *Mentzelia nuda*. In this species the floral nectar attracts pollinating bees, and the postfloral nectar attracts ants whose presence ensures enhanced seed-set. In general, if floral and extrafloral nectaries are active simultaneously they attract different visitors (Keeler 1977, 1980, 1981), and the nectars have different compositions (Baker et al. 1978). In *Mentzelia nuda* the floral and postfloral nectars have very similar composition (Keeler 1981). A somewhat analogous situation has been reported by Scott (1979) in *Banksia media,* where damage to the conflorescence by phytophagous beetles releases nectar, which attracts ants which the defend the conflorescence against further attack.

Nectar production varies widely: *Helianthella quinquenervis,* for example, secretes several microlitres of nectar per flower head per day (Inouye & Taylor 1979). Extrafloral nectars are generally aqueous sugar solutions containing a variety of additional compounds, and differ from floral nectars particularly in their amino-acid composition (Baker & Baker 1973, 1975; Baker 1978; Baker et al. 1978; Bentley 1977a; Keeler 1977, 1980a). The most extensive study is that of Baker et al. (1978), who compared the compositions of 33 extrafloral nectars with those of 248 floral nectars, including direct comparisons for 21 species which produce both. Glucose and fructose are commonly the dominant sugars in extrafloral nectars, with sucrose in lower concentration (Koptur 1979; Beckman & Stucky 1981), but *Helianthella quinquenervis* produces sucrose and no other sugars (Inouye & Taylor 1979). There can be considerable differences even between related species: *Ipomoea carnea* sepal nectar contains 36% sucrose, but *I.leptophylla* none (Keeler 1977, 1980a). Sugar concentrations cannot generally be compared with any precision as nectar dilutions are highly variable (Bentley 1976; Beckman & Stucky 1981). Amino-acids recorded in extrafloral nectars are listed by Beckman & Stucky (1981); Baker et al. (1978) have suggested that they may confer a distinctive taste to particular nectars, as well as nutritive value, but as pointed out by Inouye & Taylor (1979), 'there have been no studies of the physiological and behavioural responses of ants to amino acids.' The amino acids of *Helianthella quinquenervis* extrafloral nectar are much higher in number and concentration (16×) than those of any other so far studied (Inouye & Taylor 1979).

Nectar is costly to the plant, since metabolic energy is used in its manufacture (Luttge 1971), and it contains metabolically useful compounds (Bentley 1977b and references cited therein). Control over the timing and rate of its production would therefore be an advantage. The anatomy and physiology of extrafloral nectaries are similar to those of floral nectaries, which do possess such metabolic control, so it might also be anticipated for extrafloral nectaries (Bentley 1977b). EFN's associated with floral structures are active only during flowering (e.g. Scott 1981), but even foliar nectaries show seasonal variation in nectar production. Those of *Acacia saligna* are most active in autumn, less so in winter and summer (Majer 1979). Maximal production and sugar concentration of *Acacia longifolia* nectar coincides with flowering, when the plants are attacked by various pollen-eating beetles (Buckley ms). The extrafloral nectaries of *Prunus serotinus* are most active during the first three weeks after budbreak, when it is subject to attack by young larvae of *Malacosoma america-*

num which are still small enough to be preyed upon by *Formica obscuripes* (Tilman 1978). Hence for some species at least, such metabolic control seems to have evolved.

For the plant, the costs of EFN production are small relative to the benefits. Foliar nectar production in *Ochroma pyramidale,* for example, represents only 1% of the total energy invested in each leaf (O'Dowd 1979). How do the costs and benefits of an EFN–ant defence compare with those of alternative chemical or structural defence mechanisms? Elias (1980) points out that EFN's are more common on younger leaves, which are perhaps less able to tolerate chemical defences (Orians & Janzen 1974). Similarly Rehr et al. (1973) noted that ant-acacias lack the cyanogenic glucosides possessed by most other acacias. Inouye & Taylor (1979) suggest that *Helianthella* species such as *H.parryi* are more effectively protected by chemical defences than *H.quinquenervis* is by its ants. *H.uniflora* also lacks EFN's, but ants tend aphids, 'surrogate nectaries', on the back of the involucral bracts of this species, i.e. in precisely the same sites as the EFN's of *H.quinquenervis.* The permanent EFN's of *Opuntia acanthocarpa* assure it better protection by *Crematogaster opuntiae* than do the ephemeral nectaries of *O.phaeacantha* (Pickett & Clark 1979), but there has been no comparison with chemically protected species. The costs of extrafloral nectar are relatively slight, however, and the EFN strategy may have the added long-term advantage that whilst herbivores can rapidly evolve detoxication systems (Krieger et al. 1971; Whittaker & Feeny 1971) and behavioural modifications (Rathcke & Poole 1975) to overcome chemical and structural defences, 'ants provide a consistent defence system, relatively immune to evolutionary changes by the herbivore' (Schemske 1980). Within the genera *Inga* and *Pithecellobium,* however, there may have been a change from ant to chemical defence (Elias 1972), and in *Crescentia* and *Campsis* a change from ants to mechanical defence (Elias & Gelband 1975, 1976; Elias & Prance 1978).

Two further questions arise: firstly, is extrafloral nectar subject to theft by opportunists which do not provide plant defence; and secondly, how does the plant prevent ants consuming floral as well as extrafloral nectar? Extrafloral nectaries are visited by wasps, flies and beetles as well as ants (Putnam 1963; Keeler 1978, 1981) but these are deterred by ants guarding their food supply: this, of course, is the mechanism by which the plant gains its protection. It appears that though some floral nectars contain phenolics and alkaloids (Baker & Baker 1975; Janzen 1977; Baker 1978), they are still palatable to ants (Feinsinger & Swarm 1978; Schubert & Andersen 1978; Rico-Gray 1980). Ant access to floral nectar is prevented by mechanical barriers or by chemical deterrents in floral tissue (Guerrant & Fiedler 1981). This will be considered in more detail in connection with the scarcity of ant pollination.

Extrafloral nectar is generally not the only food of the ants involved in these interaction (e.g. Majer 1979), but its aggressive defence indicates that it is a prized resource at least. In most cases the nectar is the only reward provided by the plant (Schemske 1980) though as noted earlier some ants also live on or in the plants concerned, and their defence is presumably augmented in such cases. Some extrafloral nectars have a relatively high amino-acid content and may provide a complete diet, but in general they are rich only in sugars and would require supplementation with, e.g., arthropod prey. This is likely to be readily available in the form of the phytophagous insects attacking the plant. Evolutionarily, therefore, no great morphological, physiological or behavioural modifications have been required for the ants to participate in the interactions, though some ant species do now have very well-defined behavioural patterns associated with the interactions.

Food bodies and domatia

Various tropical plants produce specialized food bodies, analogous to extrafloral nectaries in that they attract ants which defend the plant. Besides EFN's, there are at least five classes of plant structure which serve as food resources for ants, as summarized below.

Plant	Ant	Food body	Material
Acacia	*Pseudomyrmex*	Beltian	protein and lipid
Cecropia	*Azteca*	Mullerian	glycogen
Macaranga	*Crematogaster*	Beccarian	lipid
Ochroma &c.	various	pearl	lipid
Piper	*Pheidole*	–	protein and lipid

In many cases the ants nest in or on the plant in addition to gaining a food supply. The ants defend the plants against insect and mammalian herbivores and against encroaching plant competitors, particularly vines. In the past it has been argued that the associations are fortuitous or exploitative (Bailey 1922), but the evidence for mutualistic interactions is now overwhelming.

A number of neotropical *Acacia* species possess specialized swollen thorns inhabited by ants which defend the trees against phytophagous insects, browsing mammals, and encroaching vines (Janzen 1965, 1966, 1967a, b, 1974). The acacia leaves also possess extrafloral nectaries and produce specialized food bodies, Beltian bodies, which are collected and eaten by ants (Belt 1874; Darwin 1877; Janzen 1965 et seq.). The interaction between *Acacia cornigera* and *Pseudomyrmex ferruginea* was the first ant–plant interaction experimentally demonstrated to be mutualistic (Janzen 1965), and the relation between the swollen-thorn acacias and their ants is obligate in the sense that neither survives in the absence of the other (Janzen 1966). Brown (1960) described the swollen-thorn *Acacia pseudofistulosa* as inhabited by a *Crematogaster* species, but according to Janzen (1966) all the New World obligate acacia-ants are species of *Pseudomyrmex,* though not all New World *Pseudomyrmex* species are obligate acacia-ants. Those involved include *P.ferruginea, P.belti, P.spinicola, P.nigrocincta, P.satanica, P.nigropilosa* and at least three further species (Janzen 1966). The swollen-thorn acacias include *A.cornigera, A.sphaerocephala, A.hindsii, A.collinsii* and *A.melanoceras* (Janzen 1966). Founding *Pseudomyrmex* queens find unoccupied *Acacia* seedlings or suckers and install themselves in new or empty thorns, cutting an entrance hole and excavating the soft central parenchyma if necessary. They feed on extrafloral nectar, and feed their brood on the Beltian bodies. In *P.fulvescens* at least, queens produce eggs only when there are food bodies on *A.cornigera* (Janzen 1965). The colonies expand until occupying all the thorns in one or many trees: a single colony of *P.belti* in Mexico can occupy 100 trees (Janzen 1966).

The *Cecropia – Azteca* interaction is similar in that the ants inhabit hollow *Cecropia* internodes, which they enter through weak spots in the walls; feed on specialized food bodies; and defend the plant against predators and competitors (Belt 1874; Müller 1874, 1880; Barnwell 1967; Janzen 1969a; Rickson 1967a). It differs in that *Cecropia* lacks extrafloral nectaries, and in that the Mullerian bodies are particularly rich

in glycogen (Rickson 1971), though also containing lipid and protein (Janzen 1969a). Besides the 1 mm × 3 mm Mullerian bodies, the plants produce slightly larger bodies known as pearl glands (Rickson 1976a) but these do not seem to be a major food source for the ants. At least 10 *Azteca* species, notably *A.* cf.*alfari,* are involved in obligate relationships with *Cecropia* species (Wheeler 1942; Janzen 1969a). The *Cecropia* species include *C.obtusifolia, C.peltata,* and *C.sandersoniana* (Janzen 1969a).

A third class of food body is produced by *Macaranga* species in Southeast Asia (Ridley 1910; Rickson 1980). At least nine *Macaranga* species, including *M.triloba* and *M.hypoleuca,* possess hollow stems inhabited by *Crematogaster* species, and produce lipid-rich Beccarian bodies consumed by the ants, which also form coccids in the hollow stems (Rickson 1980). *Macaranga* species without ants have solid stems, but those characteristically associated with ants produce hollow stems whether inhabited or not.

Pearl bodies, also known as bead bodies, pearl glands or pearl hairs, are produced in over 50 tropical and subtropical plant genera in at least 19 different families (Meyer 1837; Penzig 1892; Wheeler 1910; Rouppert 1926; O'Dowd 1980). They are collected by a number of ants (Raciborski 1898, 1900; Ridley 1910; Wheeler 1910; Risch et al. 1977): in particular, species of *Iridomyrmex, Rhytidoponera* and *Chelaner* will collect the pearl bodies of *Ochroma pyramidale* from artificial depots (O'Dowd 1980) though *Pheidole* will not. Pearl bodies are often associated with extrafloral nectaries; on *Ochroma pyramidale* these are situated on the leaf veins and petioles, and at least 90% of saplings in the field have at least one leaf actively producing nectar at all times, keeping the ants on the plant (O'Dowd 1980).

In all the interactions described above, the domatia, food bodies and nectaries are produced by the plant whether ants are present or not. At least three tropical *Piper* species also produce food bodies which are consumed by ants (Risch et al. 1977), and those of *Piper cenocladum* are produced only when the plant is occupied by *Pheidole bicornis:* food-body production falls drastically if ants are removed and starts again if they are replaced (Risch & Rickson 1981).

At least 5 further plant genera are involved in similar interactions. An *Azteca* species defends *Cordia alliodora* in Central America (Janzen 1969);

Pseudomyrmex species defend *Triplaris, Tachigalia* and *Ateleia* in Central and South America (Ule 1906, 1907; Janzen 1969a); *Pachysima* species defend *Barteria fistulosa* in West Africa (Janzen 1972); and a *Viticicola* is reported from *Vitex* (Janzen 1966). A number of South American species of *Gerascanthus, Pterocladon, Sapium, Platymiscium, Clidemia, Leandra, Ossaea, Cordia, Hirtella, Tococa, Maieta* and *Duroia* also possess domatia inhabited by species of *Pseudomyrmex, Solenopteris, Azteca* and *Allomerus* (Ule 1906; Janzen 1969; Neto & Asakawa 1978). These interactions, however, have not been studied in such detail as those summarized above.

The food bodies differ in position, structure, development and contents. Beltian bodies form on the tips of each rachis and pinnule. Mullerian bodies form from the trichilium coating the abaxial surface of the keel-shaped petiole bases (Rickson 1976a), and the food bodies of *Piper cenocladum* form on the recurved adaxial surfaces of petiole margins (Risch & Rickson 1981). In *Macaranga triloba* the Beccarian bodies grow on the underside of the downturned clasping stipules, and in *M.hypoleuca* on the abaxial surfaces of young leaves (Rickson 1980). The pearl bodies of *Ochroma pyramidale* are produced on the leaves and stems, and their relative abundance in different parts of the crown is correlated with those of extrafloral nectaries and ants (O'Dowd 1980). *Ochroma* pearl bodies are produced only on juvenile trees (O'Dowd 1980), and *Piper cenocladum* food bodies only in the presence of *Pheidole bicornis* (Risch & Rickson 1981). The anatomy, ultrastructure and development of the various food bodies have been described by Rickson (1975, 1976a, b, 1980) and O'Dowd (1980). *Piper cenocladum* food bodies are unicellular, the other four types multicellular. *Ochroma* may produce over 100 pearl bodies per leaf, and *Piper* around 1 500 per petiole.

Beltian bodies contain mainly protein and lipid, but the *Pseudomyrmex* acacia-ants are provided with carbohydrate from the extrafloral nectaries, which produce around 1 ml nectar/tree/day, containing 40 mg glucose and fructose (Janzen 1966). *Piper cenocladum* food bodies are also rich in protein (10.7%) and lipid (22.2%), and low in carbohydrate (2.1%); the ants obtain their carbohydrate elsewhere. Beccarian bodies contain primarily lipid, together with a few large starch grains; lipid and starch together constitute over 80% of the total cytoplasmic volume (Rickson 1980). *Ochroma pyrami-*

dale pearl bodies have an energy content of 27.8 kJ/g. dry weight and contain 74.4% lipid (O'Dowd 1980): a very high value, comparable to the highest found in seeds (Earle & Jones 1962). Beccarian bodies are unusual in containing large glycogen plastids (Rickson 1971), the glycogen being identical to animal glycogen (Marshall & Rickson 1973).

The ant domatia in these defence–domicile–nutrition interactions also vary widely in origin and structure. In *Acacia* they are swollen stipular spines, with soft parenchymatous centres which must be hollowed out by the ants. In *Cecropia, Barteria* and *Macaranga* the ants live in large hollow internodes, penetrating through weak spots in the walls near the septa (Brown 1960; Janzen 1972; Rickson 1980; Neto & Asakawa 1978). In *Tococa* ants live in twin pouches at the leaf bases (Brown 1960), and in *Piper*, under the curved petiole margins and in hollowed-out stems (Risch & Rickson 1981). In *Cordia, Hirtella, Maieta* and *Duroia* they live in various structures on twigs or petiole bases (Neto & Asakawa 1978). Ants collecting *Ochroma* pearl bodies apparently do not nest on the plant (O'Dowd 1981).

In general, the ant defences are aggressive and effective. *Pseudomyrmex* and *Crematogaster* in particular possess powerful stings which they use without hesitation, and Brown (1960) has suggested that nictitating membranes and similar features of browsing ungulates may have evolved in part as a defence against ant attack. Browsers generally avoid ant-occupied acacias, but when other food is scarce white-faced monkeys *(Cebus capucinus)* will tear open *Acacia collinsii* thorns despite the stings of *Pseudomyrmex belti,* eating ants, thorn parenchyma and fruit pulp and indirectly dispersing seed (Freese 1976). *Ateles geoffroyi* behaves similarly on *Cecropia peltata,* but *Pachysima* on *Barteria fistulosa* effectively protects the trees against *Colobus satanus* (McKey 1974) and large mammals (Janzen 1972). *Macaranga* trees lacking *Crematogaster* colonies are subject to severe damage by tortricid larvae (Rickson 1980), and the ants actively remove lepidopteran eggs. *Pseudomyrmex* workers attack all other insects on the swollen-thorn acacias, and reduce herbivore damage very significantly: unoccupied acacias lose their leaves and growing tips (Janzen 1966). There are at least nine insect species which can survive ant attack, however, and which are now restricted to the swollen-thorn acacias. These include the scarabeid *Pelidnota punctulata,* whose impenetrable cuticle al-lows it to ignore the ant attacks; the saturniid *Syssiphyx mexicana,* which for some reason is not attacked; and the noctrid *Coxina hadenoides,* which is capable of throwing off attacking ants bodily (Janzen 1966).

Besides defending the plants against herbivores, the ants chew off the tips and tendrils of encroaching vines which might otherwise smother the host plants. Generally the vines are cut only where they contact inhabited trees (Janzen 1969a; Rickson 1977), but *Pseudomyrmex* workers clear a circular area 10–150 cm in diameter around the base of inhabited swollen-thorn acacias (Janzen 1966). *Pachysima* also clears a basal area 2–3 m in diameter around *Barteria* (Janzen 1972), and removes epiphyllae from its broad leaves.

The ants thus protect the plants both against insect and vertebrate herbivores and against competing plants, performing roles otherwise taken by toxic and allelopathic plant metabolites (Janzen 1969a). *Barteria fistulosa* shoot tips lack chemical deterrents, defence being provided entirely by *Pachysima* (Janzen 1972). Swollen-thorn acacias lack fibrous reinforcement in the growing tips, as the ants protect them from phytophagous insects (except *Coxina hadenoides,* which eats them three times as fast as other acacias); hence they grow faster, and this, together with protection against vines, enables them to grow in dense secondary scrub where other acacias would be overtopped and outshaded (Janzen 1966). The cleared basal circle may also act as a firebreak, around the acacias.

What are the costs to the plant? In *Ochroma pyramidale* the total energy invested in pearl bodies is approximately a quarter of that used in the production of foliar nectar, and 0.2% that involved in leaf production; the total costs of extrafloral nectar and pearl bodies amount to 1% of leaf costs (O'Dowd 1980). The costs of nectar and food bodies are augmented by those of ant domatia, but the total costs of maintaining the ants are still far less than the savings gained, though it has not yet been possible to quantify the relative costs of ant defence system as opposed to chemical or mechanical defences. Besides producing specialized structures, plants such as swollen-thorn acacias incur the additional costs of maintaining green leaves throughout the dry season, since the ants are largely dependent on them (Janzen 1965; 1966, 1967a; Keeler 1981b). These dry-season leaves are small and lack the swollen thorns, but have a full

complement of nectaries and Beltian bodies. Since most other plants are leafless at that time, these leaves incur heavy pressure from small rodents such as *Sigmodon hispidus* (Janzen 1966).

In *Cecropia* at least, the food bodies are themselves protected chemically: an additional cost, but one presumably worthwhile. Whilst the Beltian bodies of swollen-thorn acacias are easily detached from their leaflet tips, the Mullerian bodies of *Cecropia peltata* are hidden in a mat of tannin-rich trichomes which protect them from all herbivores except those such as *Crematogaster* which search for them and extract them from within the trichilium (Rickson 1967a). *Piper cenocladum,* on the other hand, cuts costs by drastic reduction in food-body production when unoccupied (Risch & Rickson 1981).

The ants benefit from a reliable high-quality food source and in some cases also a reliable high-quality nest site. Brown (1960) notes that most plants with ant-adapted domatia also possess ant-adapted food organs. Either food or nest sites may be limiting for ant populations (Brian 1965, 1978); the relative dearth of tree-nesting ants in Australia & Melanesia, for example (Wilson 1939) is perhaps due to a shortage of suitable nest sites (Brown 1960), and the success of *Oecophylla* is perhaps largely due to its habit of constructing nests by 'sewing' together the margins of living leaves, providing it with an indefinite supply of nest sites. The swollen-thorn acacias possess four main adaptations which specifically benefit the ants: enlarged stipular thorns, Beltian bodies, enlarged foliar nectaries and year-round leaf production. The hard thorns protect ants against predation by birds such as *Icterus* species (Janzen 1966). The extrafloral nectar, perhaps supplemented by small quantities of sap from thorn parenchyma, provides the ants' sole sugar source, and together with the Beltian bodies provides the ants' entire food requirements (Janzen 1966a). The Mullerian & Beccarian bodies of *Cecropia* and *Macaranga* respectively, supplemented by coccid honeydew, are also apparently sufficient food sources (Janzen 1969a; Rickson 1980) though no ants appear to subsist entirely on *Ochroma* pearl bodies (O'Dowd 1980).

How dependent are the ants on the plants and vice versa? The *Acacia*–Pseudomyrmex interaction is an obligate mutualism (Janzen 1966): though the ants do occasionally forage for nectar off their acacias (Keeler 1981b), the ants are dependent on the plants for food and nest sites, and the plants on the ants for defence against herbivores and competitors. Neither can survive without the other; each possesses special adaptations associated with the interaction; and perhaps equally important, the acacias at least have lost adaptations which would be essential to their survival in the absence of the interaction. Only *Cecropia* produces a trichilium and Mullerian bodies; the latter are required by the associated *Azteca* species to feed their larvae (Janzen 1969a), and are particularly specialized in containing animal glycogen which the other *Azteca* species, including presumably the carnivorous ancestors of those now involved in the *Cecropia* interaction, obtain from arthropod prey. At least 10 *Azteca* species apparently have obligate relationships with *Cecropia* species (Wheeler 1942; Janzen 1969a), but *Cecropia peltata* at least can survive without its ants (Janzen 1973; Rickson 1977). The dependence of *Crematogaster* on *Macaranga* and *Pheidole bicornis* on *Piper cenocladum* have not been demonstrated experimentally, but the specializations indicate a tight relationship. Interactions involving pearl bodies such as those of *Ochroma pyramidale,* however, appear to be facultative and relatively unspecialized.

Ant–plant interactions involving food bodies and domatia have evolved independently in several unrelated ant plant genera, as outlined above. The *Pseudomyrmex–Acacia* interaction also appears to have evolved at least twice (Janzen 1966); the ants fall into two groups, and the acacias into at least five. The *Acacia–Pseudomyrmex* interaction is an excellent example of co-evolution: an evolutionary change in one species in response to a change in the second, followed by a change in the second in response to that in the first (Janzen 1966a, 1980). The *Cecropia –Azteca* and *Macaranga–Crematogaster* interactions are not all obligate, since *Cecropia peltata* grows without ants in Puerto Rico (Janzen 1973). Plants on the islands between Trinidad and Guadeloupe show a progressive reduction in ant-related traits (Rickson 1977), and the related *Musanga cecropioides* is also apparently an antless *Cecropia* (Janzen & McKay 1977). Food-body production by *Piper cenocladum* is tied to *Pheidole bicornis,* but it is not clear that *Pheidole* has undergone any reciprocal changes, so the system may not be co-evolved, though it seems likely that it is.

Ant-epiphytes

A number of epiphytic plants possess tuberous structures containing cavities regularly occupied by ants. The ants nest in the cavities and deposit debris in them; the plants absorb nutrients directly through the cavity walls or through roots within the cavities. The plants generally do not produce extrafloral nectaries or food bodies. Ant-epiphytes are particularly common in oligotrophic habitats in Southeast Asia, where nutrient supply for epiphytes would otherwise be extremely low, and where dry cavities not susceptible to termites are otherwise very rare. The ant-epiphyte symbiosis has recently been reviewed in detail by Huxley (1980).

The genera involved are *Myrmecodia, Hydnophytum, Myrmedoma, Myrmephytum, Squamellaria, Dischidia* and *Lecanopteris,* all inhabited by *Iridomyrmex* species, mostly *I.cordatus* and *I.*cf.*scrutator* (Janzen 1974; Huxley 1980); *Solenopteris,* occupied by *Azteca* (Gomez 1974, 1977); and *Tillandsia,* occupied by *Crematogaster* and other ants (Benzing 1970; Benzing et al. 1976, 1978). Species of *Wittmarckia, Aechmea, Pachycentria, Schomburgkia, Diacrium* and *Marckea* also possess structures which are irregularly or possibly inhabited by ants (Bequaert 1922; Copeland 1951; Whiffin 1972; Janzen 1974; Huxley 1980). Overall, fewer ant than plant species are involved in these interactions.

The ant-inhabited domatia in different plant genera are variously derived from hypocotyls, leaves, leaf bases, rhizomes, roots and pseudobulbs (Treub 1883a, b, 1888; Beccari 1884, 1886; Groom 1893; Scott & Sargant 1893; Thistleton-Dyer 1902; Yapp 1902; Janzen 1974; Jermy & Walker 1975; Huxley 1978), and their anatomy and development have been summarized by Huxley (1980). They range from domed leaves appressed to the trunk of the host tree in *Dischidia* to convoluted chambers in *Myrmecodia* and *Hydnophytum.*

Ants nest in the chambers and deposit debris and faeces in them. Gomez (1974) suggests that the ants excavate soft parenchyma in *Solanopteris,* but Janzen (1974) and Rauh (1973) consider that the ants merely cut an entrance and penetrate internal septa. In most genera, such as *Myrmecodia* and *Hydnophytum,* the chambers are formed entirely by the plant, with small horizontal or downward pointing entrances. The domatia in these genera often have dark brown or purple internal walls.

In most ant-epiphytes the domatia contain absorptive structures. In *Dischidia* and *Solenopteris,* for example, adventitious roots grow into some of the chambers, but do not proliferate unless debris is present (Treub 1883a; Scott & Sargant 1893; Groom 1893; Janzen 1974; Jermy & Walker 1975; Huxley 1980). In *Myrmecodia, Hydnophytum* and allied genera the internal cavity surfaces are of two distinct types, differing in color, suberization and the presence of 'warts', which are lenticular structures absorbing water, stains and radio-labelled organic and inorganic tracers (Miehe 1911a, b; Janzen 1974; Huxley 1978; Rickson 1978). In some *Myrmecodia* species the two types of surface are confined to separate chambers. Similarly, adventitious roots proliferate only in the outer chambers of the 'double flasks' of *Dischidia* (Huxley 1980).

The main roots of ant-epiphytes are primarily supportive rather than absorptive structures. For the plants, the ant-epiphyte interaction provides an 'extended feeder root system' in the form of foraging ants. In contrast to vascular epiphytes in general, the ant-epiphytes are differentially abundant in forests on oligotrophic substrates: this is particularly true of *Hydnophytym* and *Dischidia* (Janzen 1974; Huxley 1978, 1980). In such habitats, nutrient supply to epiphytes other than by ants is particularly low: the interaction enables the ant-epiphytes to grow where they have little competition. The only other vascular epiphytes in these habitats are generally associated with ant-epiphytes and appear to be essentially parasitic on their nutrient supply (Janzen 1974; Huxley 1980). This emphasizes the importance of the interaction in providing nutrients for the plants. Other benefits to the plant may include seed dispersal and 'planting' in carton, e.g. for *Myrmecodia* and *Hydnophytum* (Huxley 1980), and protection against slugs in the case of *Lecanopteris mirabilis* (Jermy & Walker 1975). Kloss gibbons, *Hylobates klossii,* in Siberut, Indonesia, which spend the night in trees of *Dipterocarpus retusus,* avoid trees bearing *Myrmecodia tuberosa* (Whitten 1982), but in general the ants are not aggressive (Huxley 1980), so defence is not a major component of the interaction. Damage to the plants by phytophagous insects is low, perhaps since they possess adequate chemical defences (Janzen 1974), notably latex in the case of *Dischidia.* Carbon dioxide released into the debris chambers is still available for photosynthesis (Billings & Godfrey 1967); the inner surfaces of *Dischidia rafflesiana* cav-

ities are richly supplied with stomata (Huxley 1982). The plants might also be able to absorb gaseous ammonia (Benzing 1970; Hutchinson et al. 1972; Janzen 1974), and moisture condensing in the chambers (Miehe 1911; Whitten 1981; Huxley 1982).

The costs to the plants are two-fold. Firstly, considerable metabolic energy is required to construct domatia, the which are formed independent of ant occupation (Huxley 1980). The biomass of the domatia is often greater than that of the leaves, for example. Secondly, the weight of these domatia, particularly the heavy tuberous structures of *Mrymecodia* and *Hydnophytum,* requires additional development of the roots as holdfasts, and even so many plants fall from the tree and die (Janzen 1974; Huxley 1980).

The primary benefit to the ants is a secure dry nest site. Dry arboreal cavities are rare in the southeast Asian forests, particularly since dead trees are subject to rapid destruction by termites, and shortage of nest sites is therefore a major constraint on ant populations in such areas (Janzen 1974; Huxley 1980, 1982). This is corroborated by the very high occupancy rate of the ant-epiphyte domatia: e.g. 95% at Bako National Park in Sarawak (Janzen 1974).

The epiphytes do not provide any specialized food tissues for the ants, which are primarily carnivorous scavengers and honeydew collectors (Janzen 1974; Huxley 1978, 1982). Ants do collect seed from some ant-epiphytes, notably *Hydnophytum formicarum,* but place them in carton around the tuber rather than consuming them (Huxley 1982).

Even in the highly specialized ant-epiphyte genera *Myrmecodia* and *Hydnophytum* the interaction is not completely obligate, since individual plants survive without their debris-collecting *Iridomyrmex* species (Janzen 1974). In such circumstances they are often inhabited by other ants, though these are displaced by *Iridomyrmex.* In general, however, the ant-epiphytes are associated with ants throughout their ranges. The associations between ant and plant are relatively specific, but one ant species may occupy several plants, and vice versa (Huxley 1980).

Specializations such as complex domatia and absorptive structures indicate a heavy investment by the plant in the interaction: this investment differs widely between the different ant-epiphyte genera. As Janzen (1974) pointed out, the heavy tubers result in a high mortality as plants fall from the host branches, implying that the tubers must confer a considerable selective advantage on those that remain attached. Features such as darkened internal chamber walls and differential distribution of absorptive sites indicate co-evolution. In summary, ant–plant interactions involving provision of nest sites by the plant and of plant nutrients by the ants have arisen independently in a number of unrelated genera, though the degree of specialization involved varies considerably.

Ant-gardens

The ant-garden associations between ants and epiphytes in tropical America are amongst the most complex of ant–plant mutualisms, though not the most specialized. Ants incorporate the epiphyte seeds into their nest carton. The plant roots form an integral part of the nest framework, and obtain nutrients from ant detritus; and the ants gain food from floral and extrafloral nectar, fruit pulp and seed arils.

The gardens are constructed of carton and contain substantial amounts of humic detritus (Weber 1943; Kleinfeldt 1978). Typically they are ovoid or elongated, ranging in size from 6 cm diameter to 23 cm by 66 cm, and are divided by thin carton walls into chambers of various sizes containing the brood (Weber 1943). Besides one or more ant species, the gardens may be occupied by beetles and parasitic phorid flies (Weber 1943).

Such associations were first reported by Ule (1901, 1906) who suggested that the ants planted seed in earth carried from below so that the roots would strengthen the nest walls. He referred to them as Ameisengarten (Ule 1901) or Blumengarten (Ule 1906) and they have become known as ant-gardens: they are, of course, quite distinct from the attine fungus-gardens.

The plants involved include species of *Philodendron, Anthurium, Streptocalyx, Aechmea, Peperomia, Codonanthe, Phyllocactus, Nidularium, Ficus, Marckea, Ectozoma* and *Codonanthopsis* (Ule 1906; Wheeler 1921; Macedo & Prance 1978; Kleinfeldt 1978). Species of *Camponotus, Azteca* and *Crematogaster* are the main ants concerned. Gardens often contain more than one ant species, the 'parabiosis' of Forel (1898), Wheeler (1921) and Weber (1943); additional ant genera involved in this way include *Dolichoderus, Odontomachus, Solenopsis, Anochetus* and *Monomorium* (Mann 1912; Wheeler 1921; Macedo & Prance 1978).

The ants forage both on and off the host tree. Wheeler (1921), for example, reported commensal *Dolichoderus* and *Crematogaster* species using the same foraging columns to collect sap and honeydew respectively, and commensal *Camponotus* and *Crematogaster* 'herding jassids and membracids' and collecting extrafloral nectar. *Crematogaster longispina* in Costa Rican *Codonanthe* ant-gardens is omnivorous, its diet including floral and extrafloral nectar, homopteran honeydew, arthropods, bird droppings, fruit and *Codonanthe* fruit pulp and arils (Kleinfeldt 1978). *Crematogaster* also tends Homoptera, but only rarely and always on nearby plants rather than on *Codonanthe* itself. *Codonanthe crassifolia* possess floral and two types of extrafloral nectary; *Crematogaster* workers tend these and sometimes enclose them in carton shelters (Kleinfeldt 1978). Whether other ant-garden plants possess such nectaries does not appear to be reported.

In some of these associations the ants defend the gardens, in others not. *Crematogaster longispina*, which forms ant-gardens with *Codonanthe crassifolia* in Costa Rica, is not aggressive and ignores alticinid beetles eating *Codonanthe* leaves (Kleinfeldt 1978). Wheeler (1943) reported that *C. limata* in British Guiana was quick to attack any intruder, but not very effective; whilst *Camponotus femoratus,* with which it forms compound gardens, was slower to attack but 'for its size the fiercest and most aggressive *Camponotus*' with a painful bite and formic acid spray. Ule (1906) had also noted that *C. limata* had a painful bite, 'etwas schmerzhafter als die *Azteca –* Arten'. Hence the ants can play a defensive role in some ant-garden associations at least, as might be expected since the gardens contain their nests.

From the plant viewpoint the ants' most significant role, however, is in placing the seeds in a favourable substrate. The only experimental analysis of which I am aware is that of Kleinfeldt (1978) who showed that the growth rate of *Codonanthe crassifolia* is significantly higher for plants in ant-gardens. Ule (1901, 1905, 1906) contended that the ants plant the epiphyte seed deliberately, since some epiphytes are found only in ant-gardens, since these species often occur as widely separated single plants where other epiphytes are absent, and since the ants do collect the seed. Wheeler (1921) disagreed with this contention, and both he and Kleinfeldt (1978) found occasional ant-garden epiphytes without ants. Kleinfeldt, however, showed that *Crematogaster longispi-*

nosa encloses *Codonanthe crassifolia* fruit in carton, consumes the fruit pulp when the fruits ripen, carries the arillate seed to the nest, and adds them to the nest walls where they germinate and sprout within a few weeks. Ants also carried *Codonanthe* seed from vines at least 2 m away, but the presence of antless *Codonanthe* nearby indicated another dispersal agent, perhaps birds. Macedo & Prance (1978) stated that *Camponotus femoratus* and *Monomorium* sp. were attracted to *Codonanthe* fruit and that the seeds became attached to the ants by their sticky arils and hence carried along branches.

Construction of carton shelters around the *Codonanthe* stem nectaries at the leaf-node bases (Kleinfeldt 1978) may also benefit the plant, since adventitious roots grow into the carton at these nodes, in contrast to those on bare areas. Overall, therefore, the main benefits to the plant appear to be seed planting and root nutrition, with defence as a possible subsidiary.

The costs to the plant appear to be minimal. Only *Codonanthe crassifolia* has been studied in any detail, and there the costs are limited to the production of extrafloral nectaries, fruit pulp, and arils. This last may also result in seed dispersal by birds, so that its costs are 'shared'.

The benefits to the ants remain unquantified. In the case of *Codonanthe crassifolia,* the *Crematogaster longispina* workers forage heavily on the extrafloral nectar, so that the interaction provides an abundant food source close to the nest. This may or may not apply to other ant-garden associations. Ule (1906) suggested that the plant roots strengthen the nest, but this remains untested and Kleinfeldt (1978) notes that *Crematogaster longispina* nests lacking *Codonanthe* were not particularly susceptible to damage.

The costs to the ants are low. If diaspores are collected for pulp or arils, the seed are better used in nest construction than discarded. Since the nests are between the plant roots, any defence is primarily of the ant brood rather than the plant. Hence overall the ant-garden associations seem to benefit both ants and plants without being expensive to either.

Perhaps for this reason, ant-garden associations are not very specific. Some ant-gardens contain several ant species: e.g. the *Camponotus femoratus* – *Crematogaster limata* – *Solenopsis* sp. compound gardens described by Weber (1943). The gardens of a particular ant species may include many different plants: those of *Camponotus femoratus* in Brazil, for example, contain various combinations of at least seven epiphytes, and those of *Azteca* any of six species (Ule 1906), though only *Philodendron myrmecophilum* was recorded from both. Similarly one plant species may grow in the gardens of several different ants: *Codonanthe crassifolia,* for example, grows in the gardens of at least six ant species (Wheeler 1921; Kleinfeldt 1978). None of the associations are known to be obligate, though some ant-garden plants do not seem to have been reported growing outside the gardens. The main garden ants always build arboreal carton nests, but do not necessarily incorporate plants.

Though ant-garden associations may now involve defence, domicile, mutual nutrition, and selective seed dispersal, and are thus the most complex of ant–plant interactions, no major adaptive jumps would have been required in their evolution. Accumulations of detritus are the usual habitat for forest epiphytes, and sweet pulp, as in the fruit and arils of *Codonanthe,* is a common item of ant diets. Extrafloral nectaries are also involved in some ant-garden associations, but do not seem to be essential to the interaction.

Seed dispersal

Seed dispersal by ants falls into two categories. First is the 'accidental' dispersal of seed abandoned in granaries of seed-harvesting ants. Second is dispersal of inedible seed bearing specialized ant-attractant food bodies or elaiosomes, by ants which collect the intact diaspores, remove the food bodies in the nest and reject the seeds. The term myrmecochory as used by Sernander (1906) referred specifically to the latter and I shall use it here in that sense, though a literal translation of the word would of course embrace both. Though there is a clear distinction in principle between the two, either mechanism may disperse seed effectively, if the proportion remaining uneaten in the former case is large enough (cf. Shea et al. 1979; Ashton 1979). There is no clear separation of the ants involved: North American species such as *Veromessor pergandei* and *Pogonomyrmex californicus,* for example, also collect elaiosome diaspores (O'Dowd & Hay 1980). Berg (1975) separated ants interacting with Australian myrmecochores into elaiosome collectors, general collectors and non-collectors, but added that the behaviour of individual species depends on conditions, competitors and alternative food sources, so that the distinction is not clearcut.

Myrmecochory is a widespread phenomenon, but the syndrome of associated plant specializations differs between regions. The three main groups recognised to date are early-flowering herbs in northern hemisphere temperate mesic forest understorey (e.g. Robertson 1897; Pudlo et al. 1980); perennials in Australian dry heath and sclerophyll woodland and forest (Berg 1975; Westoby et al. 1982), and a variety of tropical plants (Ducke & Black 1954; Horvitz & Beattie 1980). There are also myrmecochores in the North American desert, however (Solbrig & Cantino 1975; O'Dowd & Hay 1980) and probably elsewhere: very little is known about the possible incidence of myrmecochory in Africa (though see Milewski or Bond 1982), and most of South America, for example. Current evidence suggests that myrmecochory is most common in sclerophyll vegetation on infertile soils (Westoby et al. 1982b; Milewski & Bond 1982; Davidson & Morton 1981a, b, ms). Approximately 1 500 Australian plant species are believed to be myrmecochores (Berg 1975), as compared to approximately 1 000 in South Africa (N.M. Bond, cited in Westoby et al. 1982b; Milewski & Bond 1982), and 300 in the rest of the world (Berg 1975).

Myrmecochores occur in a wide range of plant families: in Australia, for example, at least 87 genera in 23 families are believed to possess myrmecochorous species (Berg 1975). Northern hemisphere myrmecochores belong to genera such as *Uvularia* (Robertson 1897; Beattie & Culver 1981); *Ulex* (Weiss 1908, 1909); *Roscoea* and *Cautleya* (Nordhagen 1932a, b); *Trillium* (Gates 1940, 1941; Beattie & Culver 1981); *Sanguinaria, Hepatica* and *Viola* (Gates 1942; Culver & Beattie 1978, 1980; Beattie et al. 1979; Pudlo et al. 1980); and *Carex, Luzula* and *Claytonia* (Handel 1976, 1978a). Beattie & Culver (1981) found that almost a third of the herbaceous understorey species in West Virginia hardwood forest are ant-dispersed, including species of *Anemone, Asarum, Corydalis, Disporum* and *Jeffersonia*. Fungal spores, notably those of *Geastrum*, may also be dispersed by ants (Sunhede 1974). Early literature, reviewed by Ridley (1930) and Van der Pijl (1972), was largely inferential: more rigorous experimental analyses of the interaction are relatively recent.

The main ants involved in myrmecochory in northern hemisphere temperate communities are species of *Aphaenogaster, Formica, Lasius, Leptothorax, Myrmica* and *Tapinoma* (e.g. Culver & Beattie 1978, 1980; Beattie & Culver 1981). In wet tropical forest understorey, *Odontomachus* and *Pachycondyla* disperse the seed of *Calathea*, and *Azteca, Paratrichima* and *Pheidole* disperse *Chrysothemis friedrichsthaliana* (Lu & Mesler 1981). In Australia the main genera are perhaps *Iridomyrmex, Monomorium, Pheidole* and *Rhytidoponera,* though a number of others have been recorded collecting elaiosome diaspores.

Sernander (1906) divided northern hemisphere myrmecochores into 15 classes on the basis of the anatomical origin of the ant-attractant appendage, diaspore presentation, and other associated features. Berg (1975) recognized at least three and perhaps 6 additional types from Australia, and Drake (1981) added another. Elaiosomes are derived from a variety of tissues, e.g. raphe, pericarp or receptacle (Sernander 1906; Ridley 1930). The most detailed treatment is that of *Kennedia* and *Hardenbergia* by Berg (1979), who discussed their elaiosome ontogeny, anatomy and morphology as compared to legume elaiosomes in general.

Temperate northern hemisphere myrmecochores often exhibit a consistent 'myrmecochore syndrome' of associated characteristics. They are generally spring-flowering herbs whose seeds ripen and are released early, and often possess persistent local chlorenchymatous assimilative structures in or around the fruit (Berg 1975). Ant-dispersal in these northern hemisphere herbs is part of an overall spring-ephemeral life-history strategy (Thompson 1981), contrasting with late-maturing bird-dispersed species. There are similar contrasts in the Australian arid zone: at Fowlers Gap, N.S.W., the ornithochorous *Acacia* species possess larger seed than their myrmecochorous congeners, and orange rather than white arils. Australian myrmecochores, however, are generally woody and lack the subsidiary features listed above (Berg 1975; Drake 1981; Davidson & Morton ms; Westoby et al. 1982b), though Drake (1981) notes that approximately half the myrmechochores on Stradbroke Island, Queensland, release their seed early. Australian myrmechochores are often diplochorous, combining ballistic seed ejection with subsequent ant dispersal, and their food bodies are hard and long lasting, in contrast to the soft elaiosomes of the mesic northern hemisphere myrmecochores (Westoby et al. 1982b). The North American desert annual *Datura discolor* also has a hard and durable food body (O'Dowd & Hay 1980).

Whilst the food bodies of genera such as *Viola* and *Calathea* are rich in lipid and fully merit the term elaiosome (Culver & Beattie 1980; Horvitz & Beattie 1980), others such as *Datura discolor* are relatively low in lipid content but rich in amino acids (O'Dowd & Hay 1980). The food bodies of different species differ not only in the main storage compounds which presumably reward ant foraging, but also in the secondary compounds which stimulate it. Many insects are attracted by fatty acids (Dethier 1947), and many elaiosomes contain ricinolic acid, which elicits a 'collecting impulse' in *Lasius fuliginosus* (Bresinsky 1963). *Datura discolor* diaspore food bodies, however, lack ricinolic acid entirely: their main fatty acid is linolenate (O'Dowd & Hay 1980), which does not stimulate collecting in *L. fuliginosus*. *Viola odorata* elaiosomes contain ricinolic acid, but the compound which attracts *Aphaenogaster rudis,* the main ant involved in dispersing its seeds, is the diglyceride 1,2-diolein (Marshall et al. 1979). In the tropical genera *Codonanthe* and *Codonanthopsis,* the ants *Camponotus femoratus* and *Monomorium* sp. are attracted to the fruit by its sweet pulp, and then disperse the seed whose sticky arils attach them to the ants externally (Macedo & Prance 1978): this is not myrmecochory in the strict sense, but may be an equally effective means of seed dispersal. Elaiosomes are generally pale or white (e.g. Sernander 1906; Ridley 1930; Horvitz & Beattie 1980, Davidson & Morton ms) but that of *Monotoca scoparia* in Australia is bright orange (Berg 1975), *Acacia concurrens* bright yellow, and *Leucopogon pimelioides* bright red (Drake 1981).

The distances over which seeds are carried by ants vary greatly. Sernander (1906) quoted distances of 15–70 m, but far shorter distances seem more usual. The mean ant transport distance for *Viola* seed in Western Virginia is 0.75 m (Culver & Beattie 1980): less than the mean ballistic dispersal distance. In southeastern Australia Berg (1975) found that a small *Pheidole* species collecting the relatively large *Hovea rosmarinifolia* diaspores rarely carried them more than 1 m, whilst *Aphaenogaster longiceps* carried *Brachyloma elaphnoides* at least 1.5 m. In Western Australia, *Iridomyrmex* workers carried 46 *Acacia pulchella* seed 1.94 m in an hour (Shea et al. 1979);

in tropical Mexico, *Odontomachus* and *Pachycondyla* carried *Calathea* seed 0.1 to 10.4 m (Horvitz & Beattie 1980). In arid inland N.S.W., however, *Iridomyrmex purpureus viridaeanus* carried *Acacia bivenosa* seed up to 77 m (Davidson & Morton, in press). Quantitative data on seed removal frequencies and rates are scarce. Seed removal frequency and distance carried are related to plant density, as shown for *Sanguinaria canadensis* by Pudlo et al. (1980), and to ant nest density (e.g. Majer 1982). O'Dowd & Hay (1980) found that 50% of complete *Datura discolor* diaspores were taken in 90 minutes, whilst Berg (1975) noted that *Pheidole* workers apparently handled every seed of *Hovea rosmarinifolia*. Drake (1981) also found very rapid removal rates for six species in a dry sclerophyll forest in Queensland, Australia.

The effects of dispersal by ants can be divided broadly into effects associated with the initial seed removal and those associated with final seed placement. The relative advantages and disadvantages of these effects have rarely been quantified: according to Culver & Beattie (1980), there had previously been 'no direct documentation of the advantage of myrmecochory to plants'. Dispersal distance itself is in most cases small, but seeds are generally carried beyond the parent canopy, releasing the seedlings from parent competition (Janzen 1970; Hamilton & Hay 1977). This is not always the case, however: in *Sanguinaria canadensis,* for example, ants take seed beyond the small parent clones in undisturbed sites, but not beyond the larger clones in disturbed sites (Pudlo et al. 1980).

Whilst dispersal generally reduces sibling competition between seedlings, seeds dispersed by ants often germinate in dense clumps from middens or abandoned granaries, intensifying sibling competition (Smallwood & Culver 1979): clumping reduces seedling emergence slightly for *Viola hirta* and *V. odorata* (Culver & Beattie 1980).

Where seeds are subject to predation by rodents, ant dispersal beyond the parent canopy has the additional advantage of reducing the proportion taken by the rodents. Granivorous rodents in the North American deserts prefer large seeds, such as those of *Datura discolor,* and forage selectively under the canopy where seed density is high. The only seed to escape are those removed soon after seedfall by group-foraging ants such as *Veromessor pergandei* (O'Dowd & Hay 1980). Such escape from rodent predation is likely to be particularly important in deserts where predators are often resource-limited and depress soil seed reserves (Brown & Davidson 1977; Nelson & Chew 1977; Borchert & Jain 1978; Whitford 1978; O'Dowd & Hay 1980). Spatially heterogeneous predation is also common in temperate communities (Frankland et al. 1963; Harper et al. 1965; Bratton 1976), and in Western Virginia, all *Viola* seed not taken by ants were eaten by caterpillars, birds and small mammals (Culver & Beattie 1978). Small rodents, particularly *Pteromyscus leucopus,* normally locate and consume 25–50% of *Asarum canadense, Jeffersonia diphylla* and *Sanguinaria canadensis* seed in Western Virginia, but if ants are excluded this proportion doubles (Heithaus 1981). Seed dispersal by harvester ants may have an additional indirect advantage for *Datura discolor:* consumption of *D. discolor* food bodies by the ant probably increases ant populations, leading to increased ant predation on *Datura's* ephemeral competitors and increased competitive pressure on its rodent predators (O'Dowd & Hay 1980). This mechanism may also operate for other myrmecochores.

Once the food bodies are removed, ants generally eject myrmecochore seed on to the nest surface or deposit them in middens around the nest. Small ants may be unable to transport large smooth seeds once the elaiosome is removed, however (Berg 1975), so as long as they possess a testa too hard for the ants to crack, such seed are effectively planted in ant nests. Both middens and nests represent 'precision dispersal' to microsites very different from the soil surface in general. The possible effects of the nest environment are many and varied (Harper & White 1974; Harper 1977). Even small nests differ from the surrounding soil in particle size, moisture availability or nutrients (Lyford 1963; Briese 1982), often possessing enhanced levels of nitrogen (Lyford 1963; Davidson & Morton ms) or phosphorus (Culver & Beattie 1978; Davidson & Morton 1981b). Middens can also have improved nutrient status and texture (Golley & Gentry 1964; Rogers & Lavigne 1974). The nest surfaces of some arid-zone ants are bare and hard, providing a very unfavourable germination site, but other species loosen the soil during nest construction, increasing rainfall penetration (Davidson & Morton ms). *Iridomyrmex purpureus* clears away any plants which may germinate on its mounds (Ettershank 1971), but *Rhytidoponera* mounds represent favourable sites for plant growth and are often heavily

vegetated (Davidson & Morton 1981a, b).

Culver & Beattie (1978) concluded that selective placement in 'safe sites' (Harper et al. 1965) is the main advantage of myrmecochory to *Viola* in West Virginia. Seeds on nests benefit from reduced predation and increased nutrient supply (Culver & Beattie 1978). The rate of seedling emergence is increased significantly, from 9 to 30%, as is the length of the first adult leaves (Culver & Beattie 1980), suggesting that seedlings emerging from ant nests become the dominant individuals in the next generation. Similarly, *Carex pedunculata* seed in ant nests grow faster and are more likely to flower the following year (Handel 1976). Seed dispersal by ants also enables the quick-germinating fugitive *Carex pedunculata* to escape competition from its congeners *C. platyphylla* and *C. plantaginea* (Handel 1978). In the Australian arid zone, selective dispersal to favourable microhabitats is the most significant effect of myrmecochory: myrmecochorous species of the two chenopod genera *Sclerolaena* and *Dissacanthus* are differentially abundant on ant mounds and have poor competitive ability off the mounds, and compete with each other for ant dispersal (Davidson & Morton 1981 a, b. Microsites occupied by myrmecochores in sclerophyll scrub on the coast of new South Wales, however, are not detectably different from those occupied by non-myrmecochores (Westoby et al. 1982b).

The processes of seed handling and elaiosome removal effectively scarify the seeds of some myrmecochores; this increases seed permeability to water and nutrients (Mayer & Poljakoff-Mayber 1963) and may therefore enhance germination. Dymes (1916) considered that scarification of *Viola odorata* seed led to earlier seed germination and more vigorous root growth: Culver & Beattie (1978, 1980) also found that scarification hastened germination in *V. odorata* and *V. hirta,* but the effect was not statistically significant. In the wet tropics, germination of *Calathea* seed is increased by removal of the aril (Horvitz & Beattie 1980).

Though most ant-dispersed seeds are ejected from the nest, some remain buried. This may be a major benefit to the plant, since it provides protection from fire, high summer temperatures, and desiccation. Seed stored in granaries are of course intended for consumption by the ants, but a significant proportion are abandoned (e.g. Culver & Beattie 1978; Majer 1982), and the seed are then protected from other predators. Seeds are more likely to be abandoned by ants which shift nest site frequently: in West Virginia, for example, the mean residence time for *Aphaenogaster rudis* and *Tapinoma sessile* colonies was only 23 days (Smallwood & Culver 1979). Abandonment is not the only mechanism of seed burial: Berg (1975) reported that an Australian *Pheidole* sp. took *Hovea rosmarinifolia* diaspores into their nests but were unable to handle the hard smooth seeds once the food bodies were removed, so that the seeds remained in the nests. Ants may also bury ejected seed deliberately, as in *Calathea* (Horvitz & Beattie 1980), or place them under leaves and stones where they are buried by rain-washed debris (Berg 1975), but such mechanisms lead only to very shallow burial. Protection of buried seed from fire and high summer temperatures is likely to be particularly significant in Australia where myrmecochore seed are subject to both. Most seed of *Acacia pulchella* and *Eucalyptus microcorys* at a Western Australian site, where summer soil surface temperatures reach 45 °C and fires are frequent, are buried at 0–12 cm in ant nests, particularly those of *Rhytidoponera inornata* and a *Melophorus* sp. (Shea et al. 1979; Majer 1982). Seed of *Acacia latericola, Bossiaea aquifolium* and *Mirbelia dilataka* are also buried in ant nests, and germinate only after wildfires which produce a heat pulse sufficient to reach them (Shea et al. 1979). Burial is not necessarily an advantage to the plant even if the seeds are abandoned. Seeds may be buried too deep for seedlings to reach the surface, or the nests may be in an unfavourable substrate. Seeds buried and abandoned in the soil nests of *Aphaenogaster longiceps* on Stradbroke Island, for example, are likely to survive (Drake 1981), whilst those taken by *Rhytidoponera metallica,* which nests in wood, are unlikely to survive even if left to germinate.

Some myrmecochorous seeds never reach ant nests; if the food body becomes detached en route the seed is generally discarded and only the food body harvested. Such intermediate lodgement may be quite common; 10% of *Datura discolor* seed, for example, do not reach ant nests (O'Dowd & Hay 1980). This may be an advantage, if these intermediate sites are favourable for germination; or a disadvantage if the seeds are found by predators. Food bodies are most likely to become detached when the seed is wedged under an obstruction, and such microsites are likely to be favourable (Westoby et al. 1982b).

The initial costs of myrmecochory to the plant are three-fold: food-body manufacture, seed protection and diaspore presentation. There may also be secondary costs such as poor seed placement or intensified seedling competition (cf. Culver & Beattie 1980). Diaspore collection is worthwhile for the ants only if the food bodies are above a certain minimum size, depending on the size of the workers. Myrmecochory would therefore be inefficient for plants with very small seeds, and is indeed rare in ephemerals (Harper et al. 1970; Baker 1972). To prevent the ants eating the seed as well as the food body, the seed requires a hard impenetrable testa, or a chemical defence. The relative costs of such protection are particularly hard to assess since both may also confer protection against other seed predators, and the hard thick smooth testa of many Australian myrmecochore seeds (Berg 1975) is probably also an adaptation to fire (Majer 1982; Westoby et al. 1982b). Davidson & Morton (ms) found that ant-dispersed *Acacia* species in the Australian arid zone have considerably higher rates of seed parasitism than bird-dispersed species, suggesting a less effective chemical defence. Some myrmecochores employ specialized diaspore presentation mechanisms, but these are unlikely to be more expensive than those required for any other dispersal mechanism. The effective cost of myrmecochory to the plant depends on the relative cost of alternative seed dispersal mechanisms, and also on the 'limiting currency' in which it is paid (Westoby et al. 1982b): in sclerophyll scrub on the New South Wales coast, the food bodies of myrmecochorous diaspores comprised 5–10% of diaspore weight and energy content, but only 0.6–3% of diaspore phosphorus content. Thus myrmecochory may be a particularly efficient seed dispersal mechanism for many Australian plants growing on low-phosphorus soils, since it requires little phosphorus: but alternative dispersal mechanisms are also cheap in phosphorus terms (Westoby et al. 1982).

Many elaiosome-producing plants are diplochorous, coupling ant dispersal with another seed-dispersal mechanism. The most common syndrome is initial ballistochory followed by myrmecochory, and this is particularly frequent in Australia (Berg 1975; Westoby et al. 1982b). It is also common in West Virginian *Viola,* only 11 *Viola* species being purely myrmecochorous (Beattie & Lyons 1975; Culver & Beattie 1978, 1980). The advantage may be that seed ejection can take seed beyond the parent canopy and reduce predation, with ants subsequently placing seeds in favourable microsites. Such diplochorous systems require a compromise between ballistic efficiency and elaiosome weight, but this is not necessarily a disadvantage since seed and elaiosome densities are similar and diaspore weight is in itself no bar to ballistic efficiency. The two systems may counteract each other if ants carry thrown seed back toward the parent plant; but again, this is not necessarily a disadvantage if the primary requirement is merely to ensure collection by ants before consumption by predators. In any event, the long-term success of a seed-dispersal mechanism depends more on the fate of the most successful seeds or the seed shadow as a whole rather than on that of those 'wasted' as long as these last do not form too high a proportion of the total. Some *Acacia* species in the Australian arid zone may occasionally be dispersed first by birds and then by ants (Davidson & Morton ms) but this does not seem to be common.

The advantage of this interaction to the ants is a spatially and temporally predictable food source rich in lipids or amino acids. The cost is that of transporting the diaspores to the nests, detaching the food bodies and ejecting the seed. The efficiency of this procedure depends on the sizes of seed and food body relative to each other and to the ants. Seeds too small are energetically not worth harvesting, whilst those too large cannot be transported. Why do the ants take the whole diaspore instead of cutting off the elaiosome on the spot? It has been argued that the food bodies provide the only handle by which the ants can move the diaspore. This seems to be true in some cases (e.g. *Pheidole* and *Hovea rosmarinifolia*) but false in many others, since the ants eject discarded seed successfully: it is in any event irrelevant since the ants are concerned to harvest the food body, not the seed.

There seem to be two main possibilities. The first is that removing the food body in the nest rather than on the ground surface may reduce total aboveground time for the workers, reducing desiccation and susceptibility to robbing by competitors. In many cases neither of these is significant, however, and since the intact diaspores are much larger than the elaiosomes, it is in any event far from certain that aboveground time is reduced. The time taken to remove the food bodies is uncertain: the seeds may remain in the nests for several hours before ejection (e.g. Horvitz & Beattie 1980) but presumably only a small proportion of this time is required to detach the food bodies.

The second possibility is suggested by Handel's observation (Handel 1976) that ants carry their larvae to the diaspores, within the nest, to feed on the elaiosomes. Perhaps the ants do not detach the elaiosomes at all, but rather allow the larvae to consume them piecemeal, ejecting the seeds once the elaiosomes are finished? This might also account for the relatively long belowground time before the seeds are thrown out, but it remains to be tested by direct observation.

Other than behavioural adaptions, the ants do not require any particular specializations to harvest myrmecochore food bodies: for them the interaction is primarily a question of optimal foraging. The plants, however, generally possess a number of specialized adaptions associated with the 'myrmecochore syndrome' outlined earlier. Besides the food body and seed coat, some myrmecochores possess specialized local assimilative apparatus, and specializations for diaspore presentation. Adaptations for diaspore presentation commonly include reflexed pedicels as in *Claytonia virginica,* and short or bending culms as in *Carex umbellata* and *C. communis* respectively (Handel 1978). The tropical myrmecochore *Calathea microcephala* produces chasmogamous seed on erect pedicels which later bend down to the ground, and cleistogamous fruit at the base of the plant where they are taken by litter-foraging ants. *C. ovandensis* produces its myrmecochorous seed on erect inflorescences, but the seeds are released at maturity and are also collected from the litter layer. Elaiosome colour may also be viewed as a specialized 'cue' for ant collection: most elaiosomes are pale or white, often contrasting with black seed as in *Claytonia perfoliata* (Woodcock 1926) or *Helleborus foetidus* (Dymes 1916). Another important specialization is in the phenology of seed maturation. The myrmecochorous herbs of northern temperate forest understorey flower and set seed early, whilst bird-dispersed species set seed later (Thompson 1981): birds do not become available as seed vectors until late in the season, and the myrmecochores benefit by seeding early before overtopped by their competitors.

Specificity is low in ant–plant interactions involving seed dispersal (Culver & Beattie 1978; Davidson & Morton ms; Majer 1982; Westoby et al. 1982). The West Virginia spring ephemerals *Sanguinaria, Hepatica* and *Viola,* for example, are dispersed mainly by *Aphaenogaster, Formica, Myrmica* and *Lasius* species: the larger ants tend to take the larger diaspores and vice versa (Beattie et al. 1979), but there are no tight specific relationships, and diaspore characteristics influence only the frequency of removal by different ants. This lack of specificity buffers the interaction against environmental disturbance (Beattie et al. 1979). Beattie & Lyons (1975) suggested that the restricted ranges of some purely myrmecochorous *Viola* species might be due to dependence on particular ant species, but this remains untested. Davidson & Morton (ms) have pointed out that though several different ant species may disperse seed of a particular plant, they will do so in different ways and with differing effects.

Though rarely dependent on individual ant species, many myrmecochores are dependent on ants in general for seed dispersal. *Sanguinaria canadensis,* for example, lacks any other dispersal mechanism. The ants are less dependent, through myrmecochore food bodies may be a seasonally important food source, and diglycerides an important item of diet (Heithaus et al. 1980). In mesic myrmecochore communities the ants are generally carnivorous, in xeric areas granivorous. The food bodies of mesic myrmecophytes are usually soft and have been suggested as 'arthropod mimics' (Dymes 1916; Carroll & Janzen 1973; Horvitz & Beattie 1980); those of xeric myrmecophytes are hard and durable and can presumably be stored in the same way as seed. Overall, therefore, the plants have undergone considerably more adaptive change and are considerably more dependent than the ants; such imbalance is often the case in mutualisms (Vandermeer & Boucher 1978; Gilbert 1978; Horvitz & Beattie 1980). Beattie & Culver (1981) suggest that the herbaceous myrmecochores of forest understorey in West Virginia, U.S.A., comprise a dispersal guild.

Myrmecochory appears to have evolved independently in many different plant families and genera. The main structures involved, the food bodies, are derived from a variety of different plant tissues, but there seem to be no major evolutionary barriers to be crossed in the development of such tissues as ant attractants. Many factors provide strong selective forces for the evolution of effective seed dispersal for example parent-sibling competition (Janzen 1980) and differential seed predation (O'Dowd & Hay 1980). The mechanism is particularly clear in *Viola hirta* and *V. odorata,* if seedlings germinating on ant nests contribute more seed to the next generation, as

suggested by Culver & Beattie (1980). Detailed evolutionary pathways within individual genera have been put forward for various genera by Berg (1958, 1966, 1969, 1972, 1979), who suggested that its evolutionary origins, in Australia at least, were perhaps in the early Tertiary (Berg 1975).

Pollination

Few plants are pollinated by ants. Ant pollination seems to be largely confined to arctic-alpine tundra and tropical hot desert area, where other animal pollen vectors are scarce. *Eritrichium aretoides, Oreoxis alpina* and *Thlaspi alpestre* in the alpine tundra of Colorado, for example, are cross-pollinated predominantly by *Leptothorax canadensis* and *Formica neurufibarbis* (Petersen 1977). Nilsson (1978) found that ants were important pollinators of *Epipactis palustris* in one locality only, solitary wasps (Eumenidae) being more important elsewhere. Ants have also been implicated in the pollination of *Theobroma cacao* in Brazil (Winder 1978) and *Syzygium tierneyanum* in tropical Australia (Hopper 1980), but such cases appear rare.

Why is ant pollination rare, and why should it be less rare in the tundra? Hickman (1974) hypothesized that ant-pollinated plants should generally possess short stems, accessible nectaries, small plain sessile asynchronously blooming flowers, small sparse pollen grains, few seeds per flower, and sparse nectar, and should generally be found in impoverished floras, growing in dense stands. The three alpine-tundra species for which Petersen (1977) described ant pollination do show most of these characteristics.

Janzen (1977), noting that ants collect the extrafloral but not floral nectar of lowland tropical plants, hypothesized that floral nectars contain ant-repellent compounds. Some floral nectars do contain alkaloids and phenolics (Baker & Baker 1975; Baker 1978) but ants do visit flowers (Baker & Baker 1978; Rico-Gray 1980) though not so often as extrafloral nectaries. Feinsinger & Swarm (1978) showed that ants take floral nectar from *Erythrina poeppigiana, E. fusca,* and *Heliconia pittacorum,* but are repelled by nectar from the hawkmoth-pollinated *Hippobroma longiflora.* Schubart & Anderson (1978) found that ants readily took nectar from three lowland tropical plants if it was made accessible by perforating the corollas or placing the flowers on the ground, and concluded that floral morphology or phenology, or perhaps more efficient and rapid nectar consumption by nectiferous pollinators, provided the main barriers to floral nectar consumption by ants. The adaptive significance of toxic nectars was considered in detail by Rhoades & Bergdahl (1981), who saw them as a general device to maximize conspecific pollen transfer, and suggested that floral nectar in rare plant species should be more copious, more predictably available and more heavily defended than that in more abundant species. Most recently Guerrant & Fiedler (1981) carried out experimental feeding trials and chemical analyses of 25 species from wet and dry forests in Costa Rica. They concluded that floral nectars are generally palatable to ants even when they contain potentially deterrent phenolics, alkaloids or non-protein amino-acids. Ant access to floral nectar is prevented either by mechanical barriers in flower structure, or by chemical deterrents in floral tissue, as suggested previously by Kerner von Marilaun (1878), Stager (1931) and Van der Pijl (1955). Such defences would also protect pollen from theft by ants. The results of their nectar-selection trials conflicted in some instances with those of Feinsinger & Swarm (1978): in particular, Guerrant & Fiedler (1981) suggested that *Hippobroma longiflora* nectar is itself palatable, but can easily be contaminated by deterrents in the floral tissue, rendering it unpalatable. I suggest that this in itself might have adaptive significance, since the nectar remains palatable to pollinators, whilst ants chewing through floral tissue to reach the nectar will contaminate their own booty before they reach it: an ideal deterrent system.

Whether ant access to flowers is prevented by chemical, morphological or phenological barriers, what are the reasons for such barriers? They prevent ant theft of floral nectar, but floral nectar is primarily an attractant to pollen vectors. What disadvantages do ants have as pollinators? There are four main possibilities. The first is that because of their fixed nests and territoriality ants don't carry pollen far enough to effect cross-pollination: they visit many flowers on the same plant, but rarely cross to another. Yet winged hymenopteran pollinators regularly quarter all the flowers on one plant before proceeding to another, and the radius of the foraging area, for some ants at least, is equivalent to that of bees and wasps. The second possibility is that ants lack specificity: they will visit any nectar source irrespective of

species, whilst bees and wasps, though visiting many different species overall, generally concentrate on a single species in any given foraging flight. Yet bees and wasps with such species-specific foraging behaviour have presumably evolved from ancestors which collected nectar and pollen indiscriminately. Specificity in other ant–plant interactions shows that ants are capable of discriminating between plant species: why should specificity in pollination not have evolved for ants in the same way as for bees and wasps?

The third possibility is that ants are ineffective pollinators since their constant cleaning and grooming removes pollen from the cuticle before it can be transferred to another plant. Yet bees and wasps groom themselves similarly, and the cuticle of wasps is as smooth as that of ants. Ants lack pollen baskets; pollen baskets are designed to retain pollen rather than release it, but perhaps pollen transfer by spillage from pollen baskets so far outweighs cuticular transport that ants are ineffective pollen vectors by comparison? Perhaps the above three factors combine to make winged hymenopterans so much more efficient than ants as pollen vectors that for plants with access to both, the former are always preferable?

There is, however, a fourth possibility. Some ants at least secrete various substances which inhibit successful pollen germination and pollen-tube growth, and grooming spreads these compounds over the entire cuticle, so that any pollen grain carried by an ant comes into contact with a pollen germination inhibitor (Beattie 1981). The substances concerned apparently serve the ants as antifungal and antibacterial agents, but their effects on pollen, incidental to the ants, have prevented ant-pollination becoming common. Myrmicacin, for example, produced by *Atta sexdens,* inhibits germination of *Camelia sinensis* pollen (Iwanami et al. 1979).

Rather than effecting pollination, ants often have predatory or destructive effects on plant flowers, robbing floral nectar and rendering the flowers less attractive to pollinators. The nectar-rich inflorescences of *Grevillea eriostachya* in arid Australia, for example, are frequently overrun by nectar-robbing ants. The impact of such nectar thieves can be considerable (Wyatt 1980).

Indirect ant–arthropod–plant interactions

Besides direct interactions between ants and plants,

there are at least two classes of indirect interaction. The first is that where ants tend or prey on herbivorous arthropods inimical to the plant. The distinction between this class and the ant–plant interactions involving plant defence in return for ant nutrition or nest sites is that in this case the plants have no direct influence on ant behaviour.

Some ants tend phytophagous insects, particularly Homoptera and lepidopteran larvae, whilst others attack them. Ants tending sap-sucking Homoptera, particularly coccids, may not necessarily disadvantage the plant if in the process of guarding the coccid 'pseudonectaries' the ants defend the plants against more damaging herbivores such as Orthoptera or Lepidopteran larvae. Whilst interactions involving ant predation on phytophagous insects are essentially 3-fold, those where ants tend phytophagous insects are generally 4-fold in that the ants guard their insect 'herds' against arthropod predators and parasitoids. The ant *Camponotus maculatus* which tends the coccid *Icerya seychellarum* on Aldabra, for example, may reduce populations of the diaspid scale predator *Chilocorus nigritus* (Hill & Blackmore 1980). If the ants also guard the plants against other phytophagous insects, the interaction is 5-fold. Other many-fold interactions are less tightly coupled, such as those where ants prey generally on entomophagous arthropods, allowing phytophagous insect populations to increase. Ants may also influence insect populations by providing nest sites: populations of the cixiid leafhopper *Oliarius vicarius,* for example, increase following introduction of the red fire ant *Solenopsis invicta,* since it inhabits unused galleries in the ant mounds (Sheppard et al. 1979). By maintaining populations of sap-feeding insects which drop honeydew on the plant, ants may increase the incidence of fungi: attendance by *Anoplolepis longipes,* for example, results in up to 90% of citrus and cinnamon leaves in the Seychelles being infected by sooty mold (Haines & Haines 1978a). Ants may increase honeydew production by increasing populations of sap-feeding insects or by stimulating increased production per insect, as suggested by Hill & Blackmore (1980) for *Icerya seychellarum* tended by *Camponotus maculatus.* Ants consume honeydew, of course, but it appears that excess production generally outweighs consumption by the ants. Owen (1980) suggested that sap-feeding insects may benefit their plant host by increasing populations of microbial nitrogen fixers in soils receiving honeydew 'rain'

under the plant, but this remains untested.

Indirect ant–arthropod–plant interactions involve a wide range of ants and plants throughout temperature and tropical regions, but are apparently rare in the semi-arid and arid zones. Perhaps the most surprising location is an intertidal halophyte mudflat in the Gulf of California, where two *Iridomyrmex* species tend *Eriococcus larreae* and a *Cryptoripersia* sp. on *Suaeda californica,* and *Brachyrmyrmex depilis* tends a *Ripersia* sp. on the roots of *Monanthochloe littoralis,* in subterranean nests which are inundated by over a metre of seawater for up to 3.5 hours, 149 times a year (Yensen et al. 1980).

Insects tended include coccids (Burns 1973; Janzen 1974b; Stout 1979; Hill & Blackmore 1980); aphids (e.g. Wood-Baker 1977; Addicott 1978; Dash 1978; Ebbers & Barrows 1980; McLain 1980; Skinner & Whittaker 1981); membracids (Way 1963; Kitching & Filshie 1974; Cookson & New 1980; Skinner & Whittaker 1981; Buckley ms); psyllids (Skinner & Whittaker 1981); and Lepidopteran larvae (Olive 1978; Schremmer 1978a; Pierce & Pearson 1981; Pierce et al. ms).

Ants also prey on scale insects (Fol'Kina 1978), aphids (Morrill 1978; Skinner 1981; Skinner & Whittaker 1981), Lepidoptera (Girling 1978; Laine & Niemela 1979; Tilman 1980; Skinner & Whittaker 1981), weevils (Morrill 1978; Sterling 1978; Jones & Sterling 1979) and a wide variety of other arthropods.

The quantitative significance of these ant-arthropod interactions can be considerable. Attendance by *Dolichoderus taschenbergi,* for example, increases the survival rate of tuliptree scales in southwestern Pennsylvania from 8.2 to 46.8% (Burns 1973). Populations of *Ceroplastes rubens* tended by *Anoplolepis longipes* on cinnamon and citrus in the Seychelles are up to twice the size of untended populations (Haines & Haines 1978a, b), and populations of *Periphyllus testudinaceus* in limestone woodland in the U.K. also increase significantly when tended by *Formica rufa* (Skinner & Whittaker 1981). The quantitative effect on the host plant was not ascertained in any of these cases. Stout (1979) suggested that *Ocotea* plants in Costa Rican wet lowland rainforest benefited more from defence by a *Myrmelachista* species than they lost to the coccids *Dysmicoccus brevipes* and *D. cryptus* tended by the ants, but this remains untested. *Crematogaster subdentata* completely exterminated one population of the San Jose scale *Quadraspidiotus perniciosus* on pear trees

(Fol'Kina 1978), but the effect on the pear trees was not quantified. Predation by *Formica aquilonia* on larvae of the geometrid *Oporinia autumnata* in Finnish Lapland, however, allowed the survival of 'green islands' of the host tree *Betula pubescens* ssp. *tortuosa*, 15–20 m in diameter, around the *Formica* nests when remaining trees were defoliated by an *Oporinia* outbreak (Laine & Niemela 1981). Similarly, predation by *Formica rufa* on aphids and lepidopteran larvae in limestone woodland caused a significant reduction in defoliation, from 8% to 1% (Skinner & Whittaker 1981). *Oecophylla smaragdina* and *Anoplolepis longipes* protect *Cocos nucifera* against the coreid *Amplypelta cocophaga* in the Solomon Islands, but *Iridomyrmex cordatus* and *Pheidole megacephala* do not (Greenslade 1971). The red fire ant, *Solenopsis invicta,* reduces midsummer damage to cotton by the boll weevil *Anthonomus grandis* in Texas (Sterling 1978; Jones & Sterling 1979). Mean consumption of *Anthonomus* larvae ranged from 22% to 66% (Sterling 1978); the frequency of weevil-damaged quadrats was always less than 17% in ant-inhabited plots, but up to 39% in control plots where *Solenopsis* had been killed out with mirex (Jones & Sterling 1979). *S. invicta* also preys on the alfalfa weevil *Hypera postica* and the pea aphid *Acyrthosiphon pisum* without damaging alfalfa (Morrill 1978), but it damages corn (Glancey 1979) and takes seed of *Paspalum dilatatum,* attracted by carbohydrates in ergot fungi (Vinson 1972). Sterling et al. (1979) found that *S. invicta* did not cause significant population reduction in any of 47 predatory arthropods in Texan cottonfields, so that local introductions to control *Anthonomus grandis* would be unlikely to produce secondary population increases in other phytophagous insects. Suppression of *S. invicta* by mirex application in *Cynodon dactylon* pastures, however, increases populations of some arthropods and reduces those of others (Howard & Oliver 1978). Ants do not necessarily reduce mortality of tended arthropods: ants attending the coccid *Pulvinarius mesembryanthemi* in Western Australia, for example, do not reduce predation by the coccinellid *Cryptolaemus montrouserieri* (Majer 1982), and though the ants do benefit the coccids by removing the honeydew they exude and hence reducing sooty mould growth, the relation is facultative as many scale populations are not attended by ants. Again, the mealybug *Pseudococcus macrozamiae* is attended by ants, but the relation is apparently of negligible im-portance to the mealybugs (Majer 1982).

Even where ants produce significant changes in populations of other arthropods, the effects on the host plant are not necessarily significant. Though ants take most eggs and small larvae of the stem borer *Eldana saccharina* on maize and sugarcane in Uganda, remaining individuals cause considerable damage (Girling 1978). Similarly, predation by *Myrmica americana, Pheidole bicarinata* and *Lasius neoniger* on the eggs of *Diabrotica virgifera,* the western corn rootworm, has no significant impact on the plant host (Ballard & Mayo 1979).

Specificity varies widely in both the ant–arthropod and arthropod–plant components of these interactions. Some lepidopteran larvae, for example, are restricted to single host plants (Common & Waterhouse 1972; Symon 1981) whilst many of the scales and aphids occupy a broad range of hosts. Ebbers & Barrows (1980), observing 3 aphid–ant associations for 10 consecutive days each, found that 81% of 103 marked ants visited only 1 of many conspecific aphid groups on particular plants; whereas Dash (1978) found at least 8 ant species attending the maize aphid *Rhopalosiphum maidis* in India. Some ants tend one coccid and eat another: *Formica rufa,* for example, tends *Periphyllus testudinaceus* but preys on *Drepanosiphum platanoidis* (Skinner & Whittaker 1981).

Costs and benefits to ants and plants vary widely between the different types of indirect ant–arthropod-plants interaction, as noted above. If ants reduce populations of phytophagous insects, the plants will benefit; if ants guard such insects, increasing their populations or feeding rate without defending the plant, the plant will suffer. If ants tend sap-sucking insects but in the process guard the plant against potentially more damaging herbivores, the plant may gain overall, as suggested for *Helianthella uniflora* by Inouye & Taylor (1979) and for *Ocotea* by Stout (1979). Ants may transport their homopteran stock to the most vigorously growing parts of the plant (Wheeler 1910; Hough 1922; Way 1963; Rosengren 1971; Carroll & Janzen 1973), which may disadvantage the plant. *Formica* apparently produces an overall positive effect on European pine plantation production, however, despite tending large aphid populations (Wellenstein 1952; Gosswald 1954, 1958; Carroll & Janzen 1973). Recent experiments near Canberra, Australia showed that in the case of an *Acacia* species with extrafloral nectaries, normally

138

guarded by an *Iridomyrmex* species, the membracid *Sextius virescens* interferes with the ant defence by attracting ants away from the nectaries, so that overall the plant suffers from the interaction (Buckley ms). A similar effect may occur for *Acacia* species occupied by the ant-attended lycaenid *Jalmenus evagoras* (Pierce et al. ms), but this has not yet been tested.

Neither *Sextius* nor *Jalmenus* occupies all the host plants in any given area, but the reason is not clear. Do the ovipositing females select individual host plants on the basis of chemical differences or ant attendance? In the case of the lycaenid *Ogyris amaryllis,* whose larvae feed on *Amyema* species in South Australia, the ovipositing females lay almost all their eggs on ant-attended plants even if these are nutritionally inferior in terms of larval growth (Atsatt 1981). *Jalmenus evagoras,* attended by an *Iridomyrmex* in dry sclerophyll eucalypt forest in the Australian Capital Territory, consistently lays its eggs on the same individual *Acacia* plants from year to year (McCubbin 1971; Common & Waterhouse 1972; Kitching et al. 1978; Kitching 1981; Pierce et al. ms). A few eggs are laid on neighbouring individuals, and since the occupied plants are generally young the lycaenids must effectively change host plants every few years. *Sextius virescens* moved from ant-attended to unoccupied plants are found and tended by ants within a few hours, so ant attendance seems unlikely to determine host choice by gravid females, particularly since their dispersal seems to be largely passive, the winged adults being blown by the wind rather than actively choosing their flight direction. Since the ant–arthropod interaction may influence host plant fitness, the selection of individual host plants by the mobile adults of these ant-attended Lepidoptera and Homoptera is important to the plant population, and remains one of the most interesting areas for further investigation. At least one further factor may be involved. Many of the ant-attended Lepidoptera and Homoptera are subject to attack by parasitoids. Ant attendance generally decreases parasitoidism (Pierce & Mead 1981; Pierce et al. ms; Buckley ms), but insect herbivores can also utilize toxins derived from the host plant as a defence against parasitoids (Price et al. 1981). The effects of host plant chemistry and ant attendance are therefore likely to interact.

The main cost to the ants, other than foraging effort, is that of guarding the honeydew-producing insect 'herds', in the interactions where this occurs. The ants gain a food source: protein from insect prey, or a rich, localized, predictable and renewable sugar source from homopteran or lepidopteran 'herds'. Sugars are themselves valuable items in an ant diet, but in addition, insect honeydews contain various amino-acids: one in particular is a precursor of formic acid and therefore particularly valuable to ants (Pierce ms, personal communication).

Honeydew is an important item of diet for many ants (Wellenstein 1952; Way 1973; Carroll & Janzen 1973; Janzen 1974b; Haines & Haines 1978a, b), but alternative, if more expensive, sugar sources are generally available. Some ants are largely dependent on Homoptera (Weber 1944; Wing 1968; Carroll & Janzen 1973), but total dependence on honeydew has not been reported. An ant defence can be important to Lepidoptera or Homoptera: some aphid species compete for the services of ants (Addicott 1978), and ant attendance is one of the advantages of group living in the membracid *Publilia concava* (McEvoy 1979). Similarly, the eggs and larvae of the lycaenids *Jalmenus evagoras* and *Glaucopsyche lygdamus,* normally tended by ants, suffer much heavier mortality from predators and parasitoids if the ants are excluded (Pierce & Mead 1981; Pierce et al. ms). The mealybugs *Dysmicoccus brevipes* and *D. cryptus* are dependent on ants for a different reason. These coccids live in small depressions inside the stems of *Ocotea pedalifolia* in lowland rainforest in Costa Rica. The stems are hollowed out by a *Myrmelachista* species which nests in the mature stems and tends the coccids in the young shoots. The coccids can only reach the plant phloem from inside the twigs, and they can only enter the twigs through the ant exit holes (Stout 1979).

In this category of interaction the plants show no specialized morphological adaptations. Ant adaptations are primarily behavioural, though the possibility of physiological adaptations to honeydew consumption have not been investigated. The adaptations are mainly on the part of the other arthropods involved. Various lepidopteran larvae possess eversible tentacles to present honeydew to ants (Schremmer 1978a; Pierce & Mead 1981; Pierce & Pearson ms). Some also possess special glands, Malicky's organs, which produce an ant-pheromone analogue which stimulates an ant defence reaction (Pierce, personal communication). This is presumably used when the larvae are attacked by predators or parasitoids. Homoptera may also possess such organs but this has not yet been investigated. The anal whips of membracids also appear to be an adaptation to the ant interaction (Kitching & Filshie 1974). Other insects possess adaptations to prevent predation by ants. These are too numerous to list, but the defensive secretions of some beetle larvae are perhaps particularly noteworthy. *Chrysomela vigintipunctata* (Coleoptera; Chrysomelidae) on willow secretes salicaldehyde and benzaldehyde, *C. populi* on poplar salicaldehyde only, and *Gastrolina depressa* on walnut secretes juglone (5-hydroxy–1,4,–benzoquinone): each of these repels *Lasius niger* (Matsuda & Sugawara 1980).

One particularly significant aspect of these ant–arthropod–plant interactions is that similar ant-tended life styles have evolved in such a wide taxonomic range of insects. The selective advantage of an ant defence seems likely to have been the main evolutionary cause of these interactions. Nor need the defence necessarily be by ants: the membracids *Tritropidea alticollum* and *Enchenopa* sp., for example, inhabiting a *Vismia* sp. in Colombia, are guarded against ants by the wasp *Parachartergus richardsi* (Schremmer 1978b).

If these interactions are damaging to the host plants, why have they not evolved an appropriate defence? Present interactions, of course, are confined to those plant species which do not have chemical defences against the phytophagous insects concerned: there are of course no such interactions on plants which do have appropriate defences. Ant-tended Homoptera and Lepidoptera still occupy a wide range of plants, however. Some of these may benefit the host plant, as noted earlier, by attracting ants which guard the plant against more damaging herbivores, but this has not yet been demonstrated experimentally. If so, the interaction might produce a selective pressure preventing the evolution of chemical defences against other Lepidoptera or Homoptera, but in quantitative terms this seems unlikely. More information is needed on the quantitative significance of such sap-sucking insects to plant fitness.

Soil modification

Ant nests frequently differ in texture and fertility from undisturbed soil, with consequent effects on vegetation. These indirect effects can persist for months or years after the nests are abandoned, in contrast to direct effects such as vegetation clearance by ants in occupied nests. The role of ants in soil modification has been considered by Petal et al. (1970, 1977) and Petal (1978), and the types of ant nests described most recently by Dumpert (1981). The nests of *Pogonomyrmex californicus* in the northeastern Colorado shortgrass plains, for example, have lower bulk density, higher sand content and higher phosphorus and nitrate concentrations than control soils (Rogers & Lavigne 1974). Similarly Gentry & Stiritz (1972) found that the surface horizons of *Pogonomyrmex badius* nests in Florida old fields, enriched with ant debris, contained approximately 10 ± 1 ppm phosphorus and 17 ± 4 ppm potassium, as compared to 6 ± 0.5 ppm P and 7 ± 1 ppm K in controls. Using *Diodia teres* as a bioassay, they found significantly more plant growth on mound soils than controls. Plant species diversity was higher on the mounds, and the differences persisted for several years after the ants had abandoned the mounds, owing to gradual mineralization of nitrogen and other nutrients.

The plant cover of temperate European ant hills also differs from that of surrounding areas. In chalk grassland in the U.K., for example, some plant species are significantly more abundant on *Lasius flavus* nests, and others significantly less so (King 1977a). Nests contained approximately 6,000 seeds/m³, and pasture soils 9,000 seeds/m³ (King 1977b). The seeds of 12 plant species were at least $4 \times$ as abundant in pasture soils as in mound soils, whilst those of *Aira praecox* were much more abundant in mound soils, comprising 2/3 of the total mound seed content. The mounds continue to increase in size as long as the grasslands are not ploughed (King 1981). Beattie & Culver (1977) found distinct patterns associated with ant nests in communities dominated by juniper, but less distinct patterns in grasslands, and noted that vegetation heterogeneity associated with ant nests is generally more pronounced in the earlier stages of plant succession. As a particularly pronounced example, ant nests on the foreshores of salt lakes in the southwestern Baraba Steppe, U.S.S.R., initiate large earth hummocks which play a major role in plant succession (Pavlova et al. 1977). Hence soil modification is a second major class of indirect ant–plant interaction.

Conclusions

The range of ant–plant interactions is too wide to make any overall generalizations. They include facultative and obligate relations, specialized and unspecialized, direct and indirect. Some have major effects on the participants, others hardly any; some are mutualistic, others entirely one-sided. I have divided them into groups on the basis of the relative roles of ant and plant in each case, but the distinction between groups is not always clear-cut, and a different classification may become preferable as future interactions are described and analyzed. The coverage is probably biased toward Australia, though I have generally avoided recapitulating results and interpretations expressed in the other chapters of this book. With a few exceptions, I have not tried to point to specific gaps in current knowledge: rather, I have tried to assemble existing information into a coherent summary to which new information can be related. I am sure this summary will soon need to be updated.

Acknowledgments

I thank the authors of the preceding chapters, and others as noted in the text, for access to manuscripts before publication.

References

The references cited in this review are given in the general bibliography (next chapter) and are therefore not repeated here.

A world bibliography of ant–plant interactions

Ralf C. Buckley

This is believed to be a relatively comprehensive bibliography of literature on ant–plant interactions up to and including 1981, but additions and corrections will be welcomed: please contact Ralf Buckley at the address given in the preface.

Abe, T., 1971. On the food sharing among 4 species of ants in a sandy grassland. Part 1. Food and foraging behaviour. *Jap. J. Ecol. 20:* 219–230.

Addicott, J. F., 1978. Competition for mutualists: aphids and ants. *Can. J. Zool. 56:* 2093–2096.

Allred, D. M. & Cole, A. C., 1979. Ants from Northern Arizona and Southern Utah, USA. *Great Basin Nat. 39:* 97–102.

Amante, E., 1967. A formiga sauva *Atta capiguara,* praga das pastagens. *Biologico 33:* 113–20.

Amante, E., 1972. Influencia de algunes fatores microclimáticos sobre a formiga sauva *Atta laevigata* (F. Smith, 1958), *Atta sexdens rubropilosa* Forel, 1908, *Atta bisphaerica* Forel, 1908 e *Atta capiguara* Goncalves, 1944 (Hymenoptera, Formicidae). In: *Formigueiros Localizados no Estado de Sao Paulo.* Sao Paulo.

Andersen, A. N., 1980. *Seed Removal by Ants at a Mallee Site in Northwestern Victoria.* Unpublished Honours thesis, Monash University.

Andersen, A. N., 1982. Seed removal by ants in the mallee, this volume.

Ashton, D. H., 1979. Seed harvesting by ants in forests of *Eucalyptus regnans* F. Muell. in central Victoria. *Aust. J. Ecol. 4:* 265–277.

Atsatt, P. R., 1981. Ant-dependent food plant selection by the mistletoe butterfly *Ogyris amaryllis* (Lycaenidae). *Oecologia 48:* 60–63.

Atsatt, P. R. & O'Dowd, D. J., 1976. Plant defense guilds. *Science 193:* 24–29.

Auclair, J. L., 1963. Aphid feeding and nutrition. *Ann. Rev. Entomol. 8:* 439–490.

Auld, T. D. & Morrison, D. A., 1981a. Plant demography down under. I. Zygote demography from conception to death. *Abstr. XIII Int. Bot. Congr.,* 286. Sydney, August 1981.

Auld, T. D. & Morrison, D. A., 1981b. Plant demography down under. II. Insect interactions in seed production and dispersal. *Abstr. XIII Int. Bot. Congr.,* 286. Sydney, August 1981.

Autuori, M., 1940. Algumas obsevaceos sobre formigas cultivadoras de fungo (Hymenoptera, Formicidae). *Revista Entomologica, Rio de Janeiro 11:* 215–26.

Autuori, M., 1942. Contribucao para o conhecimento da Sauva (*Atta spp*), II. Sauveiro inicial (*Atta sexdens rubropilosa,* Forel, 1908). *Archivos do Instituto biologico, Sao Paulo 13:* 67–86.

Autuori, M., 1950. Contribucao para o conhecimento da Sauva (*Atta* spp). (V). Numero de formas aladas e reducao dos sauveiros iniciais. *Archivos do Instituto biologico, Sao Paulo 19:* 325–331.

Autuori, M., 1956. Contribucao opara o conhecimento da Sauva (*Atta* spp.). (VI). Infestacao residual da sauva. *Archivos do Instituto biologico, Sao Paulo 23:* 109–116.

Ayre, G. L., 1957. Ecological notes on *Formica subnitens* Creighton (Hymenoptera, Formicidae). *Insectes Soc. 4:* 173–176.

Ayre, G. L., 1958. Some meteorological factors affecting the foraging of *Formica subnitens* Creighton (Hymenoptera, Formicidae). *Insectes Soc. 5:* 147–157.

Ayre, G. L., 1969. Comparative studies on the behavior of three species of ants (Hymenoptera· Formicidae). II. Trail formation and group foraging. *Can. Entomol. 101:* 118–128.

Ayre, G. L., 1971. Preliminary studies on the foraging and nesting habits of *Myrmica americana* (Hymenoptera: Formicidae) in eastern Canada. *Z. Angew Entomol. 68:* 295–299.

Ayre, G. L. & Hitchon, D. E., 1968. The predation of tent caterpillars, *Malacosoma americanum* (Lepidoptera: Lasiocampidae) by ants (Hymenoptera: Formicidae). *Can. Ent. 100:* 823–826.

Ayyar, P. N. K., 1937. A new carton-building species of ant in South India, *Crematogaster dohrni artifex* Mayr. *J. Bombay Nat. Hist. Soc. 39:* 291–308.

Bailey, I. W., 1922. Notes on neotropical ant-plants. I. *Cecropia angulata* sp. nov. *Bot. Gaz. 74:* 585–621.

Bailey, I. W., 1923. Notes on neotropical ant-plants. II. *Tachigalia paniculata* Aubl. *Bot. Gaz. 75:* 27–41.

Bailey, I. W., 1924. Notes on neotropical ant-plants. III. *Cordia nodosa* Lam. *Bot. Gaz. 77:* 39–49.

Baker, H. G., 1972. Seed weight in relation to environmental conditions in California. *Ecology 53:* 997–1010.

Baker, H. G., 1978. Chemical aspects of the pollination of woody plants in the tropics. In: Tomlinson, P. B. & Zimmerman, M. (eds.), *Tropical Trees as Living Systems.*, Ch. 3. Cambridge Univ. Press, New York.

Baker, H. G. & Baker, I., 1973a. Amino acids in nectar and their evolutionary significance. *Nature 241:* 543–545.

Baker, H. G. & Baker, I., 1973b. Some anthecological aspects of the evolution of nectar-producing flowers, particularly amino acid production in nectar. In: Heywood, V. H. (ed.), *Taxonomy and Ecology,* 243–264, Academic Press, London.

Baker, H. G. & Baker, I., 1975. Studies of nectar constitution and pollinator-plant coevolution. In: Gilbert, L. E. & Raven, P. H. (eds.), *Coevolution of Animals and Plants,* 100–140. Univ. Texas Press, Austin.

Baker, H. G. & Baker, I., 1978. Ants and flowers. *Biotropica 10:* 80.

Baker, H. G. & Baker, I., 1980. Floral nectar constituents in relation to pollination type. In: Jones, C. E. & Little, J. (eds.), *Handbook of Pollination Biology,* in press.

Baker, H. G., Opler, P. A. & Baker, I., 1978. A comparison of the amino-acid complements of floral and extrafloral nectars. *Bot. Gaz. 139:* 322–332.

Ballard, J. B. & Mayo, Z. B., 1979. Predatory potential of selected ant species on eggs of Western corn rootworm. *Environ. Entomol. 8:* 575–576.

Ballard, J. B. & Pruess, K. P., 1979. Seed selection by an ant, *Pheidole bicarinata longula* (Hymenoptera, Formicidae). *J. Kans. Entomol. Soc. 52:* 550–552.

Banks, C. J. & Nixon, H. L., 1958. Effects of the ant *Lasius niger* (L.), on the feeding and excretion of the bean aphid, *Aphis fabae* Scop. *J. Exp. Biol. 35:* 703–711.

Barbier, M. & Delage, B., 1967. Le contenu des glandes pharyngiennes de la fourmi *Messor capitatus* Latr. (Hymenoptera, Formicidae). *Comptes rendus hebdomadaires des seances de l'Academie des sciences, Paris,* 264: 1520–1522.

Barnwell, F. H., 1967. Daily patterns in the activity of the arboreal ant, *Azteca alfari.* Ecology *48:* 991–993.

Baroni-Urbani, C., 1965. Sull'attivita di foraggiamento notturna del *Camponotus nylanderi. Insectes Soc. 12:* 253–264.

Baroni-Urbani, C., 1973. Simultaneous mass recruitment in exotic ponerine ants. *Proceedings, VIIth International Congress of the International Union for the Study of Social Insects, London, 10–15 September,* 1973, 12–15.

Baroni-Urbani, C. & Kannowski, P. B., 1974. Patterns of the red imported fire ant settlement of a Louisiana pasture: some demographic parameters, interspecific competition and food sharing. *Environ. Entomol. 3:* 755–760.

Barrer, P. M. & Cherrett, J. M., 1972. Some factors affecting the site and pattern of leaf-cutting activity in the ant *Atta cephalotes* (L.). *J. Entomol.,* (A) *47:* 15–27.

Barrett, C., 1928. Ant-house plants and their tenants. *Victorian Naturalist,* 133–137.

Beatley, J. C., 1974. Phenological events and their environmental triggers in Mojave Desert ecosystems. *Ecology 55:* 856–863.

Beattie, A. J., 1971. Pollination mechanisms in *Viola. New Phytol. 70:* 343–360.

Beattie, A. J., 1981. Ants and gene dispersal in flowering plants. *Abstr. XIII Int. Bot. Congr.,* 286. Sydney, August 1981.

Beattie, A. J. & Culver, D. C., 1977. Effects of the mound nests of the ant *Formica obscuripes* on the surrounding vegetation. *Am. Midl. Nat. 97:* 390–399.

Beattie, A. J. & Culver, D. C., 1981. The guild of myrmecochores in the herbaceous flora of West Virginia forests. *Ecology 62:* 107–115.

Beattie, A. J., Culver, D. C. & Pudlo, R. J., 1979. Interactions between ants and the diaspores of some common spring flowering herbs in West Virginia, USA. *Castanea 44:* 177–186.

Beattie, A. J. & Lyons, N., 1975. Seed dispersal in *Viola* (Violaceae): adaptations and strategies. *Am. J. Bot. 62:* 714–722.

Beccari, O., 1884–6. *Piante Ospitatrici. Malesia., 2,* 340. Tipografia del istitudo Sordo-Muti, Genoa.

Beck, S. D. & Reese, J. C., 1976. Insect–plant interactions: nutrition and metabolism. In: Wallace, J. & Mansell, R. (eds.), *Biochemical Interactions Between Plants and Insects. Recent Adv. Phytochem., 10:* 1–40.

Beckman, R. L. & Stucky, J. M., 1981a. Extrafloral nectaries and plant guarding in *I. pandurata* (L.). *Bioscience 31:* 50.

Beckman, R. L. & Stucky, J. M., 1981b. Extrafloral nectaries and plant guarding in *Ipomoea pandurata* (L.). G.F.W. Mey. (Convolvulaceae). *Am. J. Bot. 68:* 72–79.

Belt, T., 1874. *The Naturalist in Nicaragua.* Bumpas, London, 403pp.

Bentley, B. L., 1976. Plants bearing extrafloral nectaries and the associated ant community: interhabitat differences in the reduction of herbivore damage. *Ecology 57:* 815–820.

Bentley, B. L., 1977a. Extrafloral nectaries and protection by pugnacious bodyguards. *Ann. Rev. Ecol. Syst. 8:* 407–427.

Bentley, B. L., 1977b. The protective function of ants visiting the extrafloral nectaries of *Bixa orellana* (Bixaceae). *J. Ecol. 65:* 27–38.

Benzing, D. H., 1970. An investigation of two bromeliad myrmecophytes: *Tillandsia butzii* Mez, *T. caput-medusae* E. Morren and their ants. *Bull. Torrey. Bot. Club 97:* 109–115.

Benzing, D. H., Henderson, K., Kessel, B. & Sulak, J., 1976. The absorptive capacities of bromeliad trichomes. *Am. J. Bot. 63:* 1009–1014.

Benzing, D. H., Seemann, J. & Renfrow, A., 1978. The foliar epidermis in Tillandsioideae (Bromeliaceae) and its role in habitat selection. *Am. J. Bot. 65:* 359–365.

Bequaert, J., 1922. Ants in their diverse relations to the plant world. *Bull. Am. Mus. Nat. Hist. 45:* 333–621.

Bequaert, J., 1924. Galls that secrete honeydew. A contribution to the problem as to whether galls are altruistic adaptions. *Bull. Brooklyn Entomol. Soc. 19:* 101–124.

Berg, R. Y., 1959. Seed dispersal, morphology, and taxonomic position of *Scoliopus,* Liliaceae. *Skr. norske Vidensk-Akad, 4.*

Berg, R. Y., 1966. Seed dispersal of *Dendromecon:* its ecologic, evolutionary, and taxonomic significance. *Am. J. Bot. 53:* 61–73.

Berg, R. Y., 1969. Distribution, seed dispersal and evolutionary history of *Vancouveria* (Berberidaceae). *Abstr. XI Int. Bot. Cong., 13.* Seattle, Wash., USA.

Berg, R. Y., 1972. Dispersal ecology of *Vancouveria* (Berberidaceae). *Am. J. Bot. 59:* 109–122.

Berg, R. Y., 1975. Myrmecochorous plants in Australia and their dispersal by ants. *Aust. J. Bot. 23:* 475–508.

Berg, R. Y., 1979. Legume seed and myrmecochorous dispersal in *Kennedia* and *Hardenbergia* (Fabaceae) with a remark on the Durian theory. *Norw. J. Bot. 26:* 229–254.

Berg, R. Y., 1981. The role of ants in seed dispersal in Australian lowland heathland. In: Specht, R. L. (ed.), *Heathlands and Related Shrublands: Analytical Studies,* 41–50. Elsevier, Amsterdam.

Bernstein, R. A., 1971. *The Ecology of Ants in the Mojave Desert: their Interspecific Relationships, Resource Utilization, and Diversity.* Unpubl. Ph.D. thesis, Univ. Calif., Los Angeles.

Bernstein, R. A., 1974. Seasonal food abundance and foraging activity in some desert ants. *Am. Nat. 108:* 490–498.

Bernstein, R. A., 1975. Foraging strategies of ants in response to variable food density. *Ecology 56:* 213–219.

Bernstein, R. A., 1976. The adaptive value of polymorphism in an alpine ant, *Formica neorufibarbis gelida* Wheeler. *Psyche 83:* 180–184.

Bernstein, R. A., 1979a. Evolution of niche breadth in populations of ants. *Am. Nat. 114:* 533–544.

Bernstein, R. A., 1979b. Schedules of foraging activity in species of ants. *J. Anim. Ecol. 48:* 921–930.

Bernstein, R. A. & Gobbel, M., 1979. Partitioning of space in communities of ants. *J. Anim. Ecol. 48:* 931–942.

Billings, W. D. & Godfrey, P. J., 1967. Photosynthetic utilization of internal carbon dioxide by hollow-stemmed plants. *Science 158:* 121–123.

Blatter, E., 1928. Myrmecosymbiosis in the Indo-Malayan flora. *J. Indian Bot. Soc. 7:* 176–185.

Blom, P. E. & Clark, W. H., 1980. Observations of ants (Hymenoptera, Formicidae) visiting extrafloral nectaries of the barrel cactus *Ferocactus gracilis* (Cactaceae) in Baja Caifornia, Mexico. *Southwest Nat. 25:* 181–196.

Bohart, G. E. & Knowlton, G. F., 1953. Notes on food habits of the western harvester ant (Hymenoptera, Formicidae). *Proc. Entomol. Soc. Wash. 55:* 151–153.

Boomsma, J. J. & de Vries, A., 1980. Ant species distribution in a sandy coastal plain. *Ecol. Entomol. 5:* 189–204.

Borch, C., 1928. Life histories of some Victorian lycaenids. *Vict. Nat. 45:* 188–193.

Bowers, M. A. & Porter, S. D., 1981. Effect of foraging distance on water content of substrates harvested by *Atta columbica* (Guerin). *Ecology 62:* 273–275.

Box, I. S., 1960. Notes on the harvester ant, *Pogonomyrmex barbatus* var. *molefacieus,* in south Texas. *Ecology 41:* 381–382.

Boyd, N. D. & Martin, M. M., 1975. Faecal proteinases of the fungus growing ant, *Atta texana:* their fungal origin and ecological significance. *J. Insect Physiol. 21:* 1815–1820.

Bradley, G. A. & Hinks, J. D., 1968. Ants, aphids, and Jack pine in Manitoba. *Can. Entomol. 100:* 40–50.

Brantjes, N. B., 1981. Ant, bee and fly pollination in *Epipactis palustris* (Orchidaceae). *Acta. Bot. Neerl. 30:* 59–68.

Breedlove, D. E. & Ehrlich, P. R., 1968. Plant-herbivore coevolution: lupines and lycaenids. *Science 162:* 671–672.

Bresinsky, A., 1963. Bau, Entwicklungsgeschichte und Inhaltsstoffe der Elaiosomen. Studien zur myrmekochoren Verbreitung von Samen und Bruchten. *Bibliotheca Bot. 126:* 1–54.

Brian, M. V., 1955. Food collection by a Scottish ant community. *J. Anim. Ecol. 24:* 336–351.

Brian, M. V., 1964. Ant distribution in a southern English heath. *J. Anim. Ecol. 33:* 451–461.

Brian, M. V., 1965. *Social Insect Populations.* Academic Press, London.

Brian, M. V., 1971. Ants and termites. In: Phillipson, J. (ed), *Methods of study in quantitative soil ecology: population, production and energy flow,* IBP Handbook No. 18, pp. 247–261. Blackwell Scientific Publications, Oxford.

Brian, M. V. (ed), 1978. *Production Ecology of Ants and Termites.* Cambridge U.P., Cambridge, 399pp.

Brian, M. V., 1979. Habitat differences in sexual production by two co-existent ants. *J. Anim. Ecol. 48:* 943–954.

Brian, M. V., Clarke, R. T. & Jones, R. M., 1981. A numerical model of an ant society. *J. Anim. Ecol. 50:* 387–405.

Brian, M. V., Hibble, J. & Stradling, D. J., 1965. Ant pattern and density in a southern English heath. *J. Anim. Ecol. 34:* 545–555.

Brian, M. V., Mountford, M. D., Abbott, A. & Vincent, S., 1976. The changes in ant species distribution during ten years postfire regeneration of a heath. *J. Anim. Ecol. 45:* 115–133.

Briese, D. T., 1974. *Ecological Studies of an Ant Community in a Semi-arid Habitat (with Emphasis on Seed-harvesting Species).* Ph. D. Thesis, Australian National University.

Briese, D. T., 1982a. The effect of ants on the soil of a semi-arid saltbush habitat, *Insectes Soc. 29:* in press.

Briese, D. T., 1982b. Relationship between the seed-harvesting ants and the plant community in a semi-arid environment. This volume.

Briese, D. T. & Macauley, B. J., 1977. Physical structure of an ant community in semi-arid Australia. *Aust. J. Ecol. 2:* 107–120.

Briese, D. T. & Macauley, B. J., 1980. Temporal structure of an ant community in semi-arid Australia. *Aust. J. Ecol. 5:* 121–134.

Briese, D. T. & Macauley, B. J., 1981. Food collection within an ant community in semi-arid Australia, with special reference to seed harvesters. *Aust. J. Ecol. 6:* 1–19.

Brough, E. J., 1976. Notes on the ecology of an Australian desert species of *Calomyrmex* (Hymenoptera, Formicidae). *J. Aust. Entomol. Soc. 15:* 339–346.

Brown, E. S., 1959a. Immature nutfall of coconuts in the Solomon Islands. I. Distribution of nutfall in relation to that of *Amblypelta* and of certain species of ants. *Bull. Entomol. Res. 50:* 97–134.

Brown, E. S., 1959b. Immature nutfall of coconuts in the Solomon Islands. II – Changes in ant populations, and their relation to vegetation. *Bull. Ent. Res. 50:* 523–558.

Brown, J. H., 1973. Species diversity of seed-eating desert rodents in sand dune habitats. *Ecology 54:* 775–787.

Brown, J. H., 1975. Geographical ecology of desert rodents. In: Cody, M. L. & Diamond, J. M. (eds.), *Ecology and Evolution of Communities,* 315–341. Belknap Press, Cambridge, Mass.

Brown, J. H. & Davidson, D. W., 1977. Competition between seed-eating rodents and ants in desert ecosystems. *Science 196:* 880–882.

Brown, J. H., Davidson, D. W. & Reichman, O. J., 1979. An experimental study of competition between seed-eating desert rodents and ants. *Am. Zool. 19:* 1129–1143.

Brown, J. H., Grover, J. J., Davidson, D. W. & Lieberman, G. A., 1975. A preliminary study of seed predation in desert and montane habitats. *Ecology 56:* 987–992.

Brown, J. H. & Lieberman, G. A., 1973. Resource utilization and coexistence of seed-eating desert rodents in sand dune habitats. *Ecology 54:* 788–797.

Brown, J. H., Reichman, O. J. & Davidson, D. W., 1979. Granivory in desert ecosystems. *Ann. Rev. Ecol. Syst. 10:* 201–227.

Brown, W. L., 1960. Ants, acacias and browsing mammals. *Ecology 41:* 587–592.

Brown, W. L. & Taylor, R. W., 1970. Hymenoptera: superfamily Formicoidea. In: CSIRO, *The Insects of Australia,* 951–959. Melbourne University Press, Melbourne.

Brues, C. T., 1924. The specificity of food-plants in the evolution of phytophagous insects. *Am. Nat. 58:* 127–144.

Brues, C. T., 1972. Insects, food and ecology. Dover Publ, New York.

Bucher, E. H. & Zuccardi, R. B., 1967. Significacion de los hormigueros de *Atta vollenweideri* Forel como alteradores del suelo en la provincia de Tucuman. *Acta zoologica lilloana 23:* 83–96.

Buckley, R. C. ms. Homopteran short-circuits acacias's ant defence.

Bullock, S. H., 1974. Seed dispersal of *Dendromecon* by the seed predator *Pogonomyrmex. Madrono 22:* 378–379.

Bullock, S. H., 1978. Plant abundance and distribution in relation to types of seed dispersal in chaparral. *Madrono 25:* 104–105.

Burger, W. C., 1972. Evolutionary trends in the Central American species of *Piper* (Piperaceae). *Brittonia 24:* 356–362.

Burgett, D. M. & Young, R. G., 1974. Lipid storage by honey ant repletes. *Ann. Entomol. Soc. Amer. 67:* 743–744.

Burns, A. N., 1931. Habits and life histories of some Victorian lycaenid butterflies. *Vict. Nat. 48:* 129–136.

Burns, D. P., 1973. The foraging and tending behaviour of *Dolichoderus taschenbergi* (Hymenoptera, Formicidae). *Can. Entomol. 105:* 97–104.

Butler, J. F., Camino, M. L. & Perez, T. O., 1978. *Boophilus microplus* and the fire ant *Solenopsis germinata. Recent Advances in Acarology 1:* 469–472.

Cammaerts, M.-C. & Cammaerts, R., 1980. Food recruitment strategies of the ants *Myrmica sabuleti* and *Myrmica ruginodis. Behav. Processes. 5:* 251–270.

Campbell, M. H., 1966. Theft by harvesting ants of pasture seed broadcast on unploughed land. *Aust. J. Exp. Agric. Anim. Husb. 6:* 334–338.

Campbell, M. H., 1982. Restricting losses of aerially-sown seed due to seed-harvesting ants, this volume.

Campbell, M. W. & Gilmour, A. R., 1979. Reducing losses of surface-sown seed due to harvesting ants. *Aust. J. Exp. Agric. Anim. Husb. 19:* 706–711.

Campbell, M. H. & Swain, F. G., 1973. Factors causing losses during the establishment of surface-sown pastures. *J. Range Manage. 26:* 355.

Campbell, T. E.. 1970. Pine direct seeding not imperiled by Florida harvester ants. *US For. Serv. Res. Note 108:* 1–3.

Campbell, T. E., 1974. Red imported fire ant: a predator of direct – seeded longleaf pine. *US For. Serv. Res Note. 179:* 1–3.

Carne, W. M., 1913. The secretion of nectar by extrafloral glands in the genus *Acacia* (wattles). *Aust. Nat. 2:* 198–199.

Carroll, C. R., 1979. A comparative study of 2 ant faunas: the stem-nesting ant communities of Liberia, West Africa, and Costa-Rica, Central America. *Am. Nat. 113:* 551–562.

Carroll, C. R. & Janzen, D. H., 1973. Ecology of foraging by ants. *Ann. Rev. Ecol. Syst. 4:* 231–257.

Carroll, J. F., Kimbrough, J. W. & Whitcomb, W. H., 1981. Mycophagy by *Aphaenogaster* spp. (Hymenoptera, Formicidae). *Proc. Entomol. Soc. Wash. 83:* 326–311.

Cates, R. G. & Orians, G. H., 1975. Successional status and the palatability of plants to generalized herbivores. *Ecology 56:* 410–418.

Champ, B. R. & Sillar, D. I., 1961. Pellet your buffel seed and thwart ants. *Queensland, Agric. J. 87:* 583.

Champ, B. R., Sillar, D. I. & Lavery, H. J., 1961. Seed-harvesting ant control in the Cloncurry district. *Queensl. J. Agric. Sci. 18:* 257–260.

Cherrett, J. M., 1968. The foraging behavior of *Atta cephalotes* (Hymenoptera, Formicidae). I. Foraging pattern and plant species attacked in tropical rain forest. *J. Anim. Ecol. 37:* 387–403.

Cherrett, J. M., 1972a. Chemical aspects of plant attack by leafcutting ants. In: J. B. Harborne. (ed.), *Phytochemical Ecology,* 13–24. Academic Press, London.

Cherrett, J. M., 1972b. Some factors involved in the selection of vegetable substrate by *Atta cephalotes* (L.) (Hymenoptera, Formicidae) in tropical rain forest. *J. Anim. Ecol. 41:* 647–660.

Cherrett, J. M., Pollard, G. V. & Turner, J. A., 1974. Preliminary observations on *Acromyrmex landolti* (For.) and *Atta laevigata* (Fr. Smith) as pasture pests in Guyana. *Trop. Agriculture 51:* 69–74.

Cherrett, J. M. & Seaforth, C. E., 1968. Phytochemical arrestants for the leaf-cutting ants, *Atta cephalotes* (L.) and *Acromyrmex octospinosus* (Reich), with some notes on the ants' response. *Bull. Entomol. Res. 59:* 615–625.

Chew, E. A. & Chew, R. M., 1980. Body size as a determinant of small-scale distributions of ants in evergreen woodland, southeastern Arizona, USA. *Insectes Soc. 27:* 189–202.

Chew, R. M., 1977. Some ecological characteristics of the ants of a desert shrub community in southeastern Arizona, USA. *Am. Midl. Nat. 98:* 33–49.

Chew, R. M. & Chew, A. E., 1970. Energy relationships of mammals of a desert shrub (*Larrea tridentata*) community. *Ecol. Monogr. 40:* 1–21.

Chew, R. M. & de Vita, J., 1980. Foraging characteristics of a desert ant assemblage: functional morphology and species separation. *J. Arid. Environ. 3:* 75–83.

Clark, J., 1934. Notes on Australian ants, with descriptions of new species and a new genus. *Mem. Nat. Mus. Vict. 8:* 5–20.

Clark, W. H. & Comanor, P. L., 1973a. The use of western harvester ant (*Pogonomyrmex occidentalis*) seed stores by rodents. *Biol. Soc. Nev. Occas. Pap. 34:* 1–6.

Clark, W. H. & Comanor, P. L., 1973b. A quantative examination of spring foraging of *Veromessor pergandei* in northern Death Valley, California, USA. (Hymenoptera, Formicidae). *Am. Midl. Nat. 90:* 467–474.

Clark, W. H. & Comanor, P. L., 1973c. Notes on the interactions between individuals of two colonies of *Pogonomyrmex occidentalis* (Hymenoptera, Formicidae). *Entomologist 106:* 277–278.

Clark, W. H. & Comanor, P. L., 1975. Removal of annual plants from the desert ecosystem by western harvester ants, *Pogonomyrmex occidentalis. Environ. Entomol. 4:* 52–56.

Clifford, H. T. & Drake, W. E., 1981. Pollination and dispersal in eastern Australian heathlands. In: Specht, R. L. (ed.), *Heathlands and Related Shrublands: Analytical Studies,* 51–60, Elsevier, Amsterdam.

Cody, M. L., Fuentes, E. G., Glanz, W., Hunt, J. H. & Moldenke, A. R., 1977. Convergent evolution in the consumer organisms of Mediterranean Chile and California. In: Mooney, H. A., (ed.) *Convergent Evolution in California and Chile: Mediterranean climate Ecosystems. I. B. P. Synthesis Series, 5:* 144–192. Dowden, Hutchinson and Ross, Stroudsberg.

Cole, A. C., 1932a. The ant, *Pogonomyrmex occidentalis* Cr., associated with plant communities. *Ohio J. Sci. 32:* 10–20.

Cole, A. C., 1932b. The relation of the ant, *Pogonomyrmex occidentalis* Cr., to its habitat. *Ohio J. Sci. 32:* 133–146.

Cole, A. C., 1934a. A brief account of aestivation and overwintering of the occident ant, *Pogonomyrmex occidentalis* Cresson, in Idaho. *Can. Ent. 66:* 193–198.

Cole, A. C., 1934b. An ecological study of the ants of the southern desert shrub region of the United States. *Ecology 9:* 388–405.

Cole, A. C., 1939. The life history of a fungus-growing ant of the Mississippi Gulf coast. *Lloydia 2:* 153–160.

Cole, A. C., 1968. *Pogonomyrmex Harvester Ants.* Univ. Tennessee Press, Knoxville, Tennessee.

Common, F. I. B. & Waterhouse, D. F., 1972. *Butterflies of Australia.* Angus and Robertson, Sydney.

Conway, J. R., 1980. Foraging behavior of the honey ant *Myrmecocystus mexicanus* in Colorado, USA. *Trans. Ill. State Acad. Sci. 72:* 81–93.

Cookson, L. & New, T. R., 1980. Observations on the biology of *Sextius virescens* (Fairmaire) (Homoptera, Membracidae) on *Acacia* in Victoria. *Aust. Ent. Mag. 7:* 4–10.

Copeland, E. B., 1951. A new genus of ferns. *Am. Fern J. 41:* 75–76.

Coyne, J. & Schneider, S., 1974. The foraging behavior of leaf-cutter ants. Organization for Tropical Studies Coursebook. *74-II,* 108–120.

Craven, E., Dix, M. W. & Michaels, G. E., 1970. Attine fungus gardens contain yeasts. *Science 169:* 184–186.

Creighton, W. S., 1950. The ants of North America. *Bull. Mus. Comp. Zool. 104:* 1–585.

Creighton, W. S., 1953. New data on the habits of the ants of the genus *Veromessor. Am. Mus. Novit. 1612:* 1–18.

Creighton, W. S., 1966. The habits of *Pheidole ridicula* Wheeler with remarks on habit patterns in the genus *Pheidole. Psyche 73:* 1–7.

Creighton, W. S. & Creighton, M. P., 1959. The habits of *Pheidole militicida* Wheeler (Hymenoptera: Formicidae). *Psyche 66:* 1–12.

Cremer, K. W., 1966. Treatment of *Eucalyptus regnans* seed to reduce losses to insects after sowing. *Aust. For. 30:* 162–174.

Crowell, K. L., 1968. Rates of competitive exclusion by the Argentine ant in Bermuda. *Ecology 49:* 551–555.

CSIRO Div. Entomol, 1970. *The Insects of Australia.* Melbourne Univ. Press, Melbourne.

Culver, D. C., 1972. A niche analysis of Colorado ants. *Ecology 53:* 126–131.

Culver, D. C., 1974. Species packing in Caribbean and north temperate ant communities. *Ecology 55:* 974–988.

Culver, D. C., 1980. The fate of *Viola* seeds dispersed by ants. *Am. J. Bot. 67:* 710–714.

Culver, D. C. & Beattie, A. J., 1978. Myrmecochory in *Viola:* dynamics of seed–ant interactions in some West Virginia species. *J. Ecol. 66:* 53–72.

Culver, D. C. & Beattie, A. J., 1980. The fate of *Viola* seeds dispersed by ants. *Am. J. Bot. 67:* 710–714.

Curtis, J. D. & Lersten, N. R., 1978. Heterophylly in *Populus grandidentata* (Salicaceae) with emphasis on resin glands and extrafloral nectaries. *Am. J. Bot. 65:* 1003–1010.

Czechowski, W., 1975. Bionomics of *Formica pressilabris* (Hymenoptera, Formicidae). *Ann. Zool. 33:* 103–125.

Czerwinski, Z., Jakubczyk, H. & Petal, J., 1971. Influence of ant hills on the meadow soils. *Pedobiologia 11:* 277–285.

Darwin, F., 1877. On the glandular bodies of *Acacia sphaerocephala* and *Cecropia peltata* serving as food bodies. *J. Linn. Soc. Lond. Bot. 15:* 398–409.

Das, G. M., 1959. Observations on the associations of ants with coccids of tea. *Bull. Entomol. Res. 50:* 437–448.

Dash, A. R., 1978. Ants attending the maize aphid. *Rhopalosiphum maidis. Prakruti Utkal. Univ. J. Sci. 12:* 139–140.

David, C. T. & Wood, D. L., 1980. Orientation to trails by a carpenter ant, *Camponotus modoc* (Hymenoptera, Formicidae) in a giant sequoia forest. *Can. Entomol., 112:* 993–1000.

Davidson, D. W., 1976. *Species Diversity and Community Organization in Desert seed-eating Ants.* Ph.D. Thesis, Univ. Utah, Salt Lake City.

Davidson, D. W., 1977a. Species diversity and community organization in desert seed-eating ants. *Ecology 58:* 711–724.

Davidson, D. W., 1977b. Foraging ecology and community organisation in seed-eating ants. *Ecology 58:* 725–737.

Davidson, D. W., 1978a. Size variability in the worker caste of a social insect (*Veromessor pergandei* Mayr) as a function of the competitive environment. *Am. Nat. 112:* 523–532.

Davidson, D. W., 1968b. Experimental tests of the optimal diet in two social insects. *Behav. Ecol. Sociobiol. 4:* 35–41.

Davidson, D. W., 1980. Some consequences of diffuse competition in a desert ant community. *Am. Nat. 92:* 92–105.

Davidson, D. W., Brown, J. H. & Inouye, R. S., 1980. Competition and the structure of granivore communities. *Bioscience 30:* 233–238.

Davidson, D. W. & Morton, S. R., 1981a. Competition for dispersal in ant-dispersed plants. *Science 213:* 1259–1261.

Davidson, D. W. & Morton, S. R., 1981b. Myrmecochory in some plants (F. Chenopodiaceae) of the Australian arid zone. *Oecologia 50:* 357–366.

Davidson, D. W. & Morton, S. R., ms. Coevolution of *Acacia* with ants and birds in the Australian arid zone. *Ecology,* in review.

Davison, E. A., 1980. *Ecological Studies of Two Species of Harvester Ants (Chelaner whitei* and *C. rothsteini) in an Arid Habitat in South Eastern Australia (Hymenoptera: Formicidae).* Ph.D. thesis, U.N.E., Armidale.

Davison, E. A., 1982. Seed utilization by harvester ants in western NSW, this volume.

De Bruyn, G. J., Goosen-De Roo, L., Hubregtse-Van den Berg,

A. I. M. & Feijen, H. R., 1972. Predation of ants by woodpeckers. *Ekol. Pol. 20:* 83–91.

De Bruyn, G. J. & Mabelis, A. A., 1972. Predation and aggression as possible regulatory mechanisms in *Formica. Ekol. Pol. 20:* 92–101.

Delpino, F., 1886. Funzione mirmecofile net regno vegetale. *Mem. R. Accad. Sci. Bologna. Serie 4. VII.* 215–323, *VII,* 601–659; *X,* 115–147.

Dethier, V. G., 1947. Chemical insect attractants and repellents. Blakiston, Philadelphia, Pennsylvania, USA.

Dethier, V. G., 1970. Chemical interactions between plants and insects. In: Sondheimer, E. & Simeone, J. B. (eds.), *Chemical Ecology.* Academic Press, New York.

Deuth, D., 1977. The function of extrafloral nectaries in *Aphelandra deppeana* Schl. & Cham. (Acanthaceae) *Brenesia 10/11:* 135–145.

De Vita, J., 1975. *Competitive and Interactive Mechanisms among Colonies of the Harvester Ant Pogonomyrmex californicus in the Mojave Desert.* Dissertation, Uni. Southern California, Los Angeles.

De Vita, J., 1979. Mechanisms of interference and foraging among colonies of the harvester ant *Pogonomyrmex californicus* in the Mojave Desert. *Ecology 60:* 729–737.

Dlusskii, G. M., Voltsit, O. V. & Sulkhanov, A. V., 1978. Organisation of group foraging in ants of the genus *Myrmica. Zool. ZH. 57:* 65–77.

Dobrzanska, J., 1966. The control of the territory by *Lasius fuliginosus. Latr. Acta Biol. Exp., Warsaw 26:* 193–213.

Dobrzanska, J., 1976. The foraging behavior of the ant *Myrmica laevinodis. Acta Neurobiol. Exp. 36:* 545–559.

Dolva, J. M. & Scott, J. K., 1982. The association between the mealybug, *Pseudococcus macrozamiae* (Homoptera), ants and the cycad, *Macrozamia reidlei* in a fire prone environment. *J. Roy. Soc. West. Aust.,* in press.

Doncaster, C. P., 1981. The spatial distribution of ants' nests on Ramsey Island, South Wales. *J. Anim. Ecol., 50:* 195–218.

Drake, W. E., 1981. Ant–seed interaction in dry sclerophyll forest on North Stradbroke Island, Queensland. *Aust. J. Bot. 29:* 293–309.

Ducke, A. & Black, G. A., 1954. Notas sobre a fitogeografica da Amazonia Brasiliera., *Bol. Tecn. Inst. Agron. N, 29.*

Du Merle, P., Jourdheuil, P., Marro, J. P. & Mazet, R., 1978. Seasonal fluctuations of the myrmecofauna, and its predatory activity in forest: interaction between 3 habitats: clearing, forest edge, forest. *Ann. Soc. Entomol. Fr. 14:* 141–158.

Du Merle, P. & Luquet, G. C., 1978. The ant and grasshopper communities of Mount Ventoux, France. Part 1. Preliminary remarks and description of the habitats studied. *Terre vie 32:* 174–160.

Dumpert, K., 1981. *The Social Biology of Ants* (transl. C. Johnson). Pitman, London.

Duncan-Weatherby, A. H., 1953. Some aspects of the biology of the mound ant *Iridomyrmex detectus* (Smith). *Aust. J. Zool. I:* 178–192.

Dutcher, J. D. & Sheppard, D. C., 1981. Predation of pecan weevil *Curculio caryae* larvae by red imported fire ants *Solenopsis invicta. J. Ga. Entomol. Soc. 16:* 210–213.

Dymes, T. A., 1916. The seed mass and dispersal of *Helleborus foetidus* Linn. *J. Linn. Soc. 43:* 433–455.

Earle, F. F. & Jones, Q., 1962. Analyses of seed samples from 113 plant families. *Econ. Bot. 16:* 221–250.

Eastop, V. F., 1973. Deductions from the present day host plants of aphids and related insects. In: Van Emden, H. F. (ed.), *Insect/plant relationships.* Wiley, New York.

Ebbers, B. C. & Barrows, E. M., 1980. Individual ants specialise on particular aphid herds (Hymenoptera, Formicidae; Homoptera, Aphididae). *Proc. Entomol. Soc. Wash. 82:* 405–407.

Echols, H. W., 1966. Assimilation and transfer of Mirex in colonies of Texas leaf-cutting ants. *J. Econ. Entomol. 59:* 1336–1338.

Edmunds, G. F. & Alstad, D. N., 1978. Coevolution in insect herbivores and conifers. *Science 199:* 941–945.

Edwards, W. H., 1978. On the larvae of *Lycaenopsis pseudargiolus* and attendant ants. *Canad. Ent. 10:* 131–136.

Elias, T. S., 1972. Morphology and anatomy of foliar nectaries of *Pithecellobium macradenium* (Leguminosae). *Bot. Gaz. 133:* 38–42.

Elias, T. S., 1979. Extrafloral nectaries – their structure and distribution In: Elias, T. S. & Bentley, B. L. (eds.), *The Biology of Nectaries.* Columbia University Press, New York.

Elias, T. S., 1980. Foliar nectaries of unusual structure in *Leonardoxa africana* (Leguminosae), an African obligate myrmecophyte. *Am. J. Bot., 67:* 423–425.

Elias, T. S. & Elias, T. S., 1976. Morphology and anatomy of floral and extrafloral nectaries in *Campsis* (Bignoniaceae). *Am. J. Bot. 63:* 1349–1353.

Elias, T. S. & Gelband, H., 1975. Nectar: its production and functions in trumpet creeper. *Science 189:* 289–291.

Elias, T. S. & Gelband, H., 1976. Morphology and anatomy of floral and extrafloral nectaries in *Campsis* (Bignoniaceae). *Am. J. Bot. 63:* 1349–1353.

Elias, T. S. & Prance, G., 1978. Nectaries on the fruit of *Crescentia* and other Bignoniaceae. *Brittonia 30:* 175–181.

Elias, T. S., Rozich, W. R. & Newcombe, L., 1975. The foliar and floral nectaries of *Turnera ulmifolia. Am. J. Bot. 62:* 570–576.

Elton, C., 1932. Territory among wood ants (*Formica rufa* L.) at Picket Hill. *J. Anim. Ecol. 1:* 69–76.

Emlen, J. M., 1966. The role of time and energy in food preference. *Am. Nat. 100:* 611–617.

Emmel, T. C., 1967. Ecology and activity of leaf-cutting ants (*Atta* sp.). In: *Advanced Zoology: Insect Ecology in the Tropics.,* 125–131. San Jose, Costa Rica: Organisation for Tropical Studies.

Erickson, J. M., 1971. The displacement of native ant species by the introduced Argentine ant, *Iridomyrmex humilis* Mayr. *Psyche 78:* 257–266.

Ettershank, G., 1971. Some aspects of the ecology and nest microclimatology of the meat ant, *Iridomyrmex purpureus* (Smith). *Proc. Roy. Soc. Aust., 84:* 137–151.

Ettershank, G. & Whitford, W. G., 1973. Oxygen consumption of two species of *Pogonomyrmex* harvester ants (Hymenoptera: Formicidae). *Comp. Biochem. Physiol., 46A,* 605–611.

Evans, H. C. & Leston, D., 1971. A ponerine ant (Hymenoptera: Formicidae) associated with Homoptera on cocoa in Ghana. *Bull. Entomol. Res., 61:* 357–362.

Faegri, K. & Van der Pijl, L., 1979. *The Principles of Pollination Ecology,* (3rd edn.). Pergamon, Oxford.

Fall, H. C., 1937. The North American species of *Nemadus* Thom., with descriptions of new species (Coleoptera, Silphidae). *J. N Y Entomol. Soc, 45:* 335–340.

Feener, D. H., 1981. Competition between ant species: outcome controlled by parasitic flies. *Science, 214:* 815–817.

Feeny, P., 1975. Biochemical evolution between plants and their insect herbivores. In: Gilbert, L. E. & Raven, P. H. (eds.), *Coevolution of Animals and Plants*, 3–19. Univ. Texas Press, Austin.

Feeny, P., 1976. Plant apparency and chemical defense. In: Wallace, J. W. & Mansell, R. L. (eds.), *Biochemical Interaction between Plants and Insects. Recent Advances in Phytochemistry, 10:* 1–40. Plenum Press, New York.

Feinsinger, P. & Swarm, L. A., 1978. How common are ant repellent nectars? *Biotropica 10:* 238–239.

Fenner, D. H., 1981. Notes on the biology of *Pheidole lamia* (Hymenoptera, Formicidae) at its type locality, Austin, Texas, USA. *J. Kans. Entomol. Soc. 54:* 269–277.

Finnegan, R. J., 1973. Diurnal foraging activity of *Formica sublucida, Formica sanguinea subnuda* and *Formica fossaceps* (Hymenoptera, Formicidae) in Quebec. *Can. Entomol. 105:* 441–444.

Fol'Kina, M. Ya., 1978. A case of mass extermination of San-Jose scale *Quadraspidiotus perniciosus* by the ant *Crematogaster subdentata. Zool. ZH., 57:* 301.

Forbes, H. O., 1880. Notes from Java. *Nature, 22:* 148.

Forel, A., 1898. La Parabiose chez les Fourmis. *Bull. Soc. Vaud. Sc. Nat. 34:* 380–384.

Fowler, H. G., 1978. Foraging trails of leaf-cutting ants. *J. N. Y. Entomol. Soc. 86:* 132–136.

Fowler, H. G., 1979. Seed predator responses. *Oecologia 41:* 361–363.

Fowler, H. G. & Cabello, L., 1978. Bionomics of *Acromyrmex lundi pubescens* in Paraguay (Hymenoptera, Formicidae). *Entomol. News 89:* 175–177.

Fowler, H. G. & Robinson, S. W., 1978. Foraging and grass selection by the grass cutting ant, *Acromyrmex landolti fracticornix* (Hymenoptera, Formicidae) in habitats of introduced forage grasses in Paraguay. *Bull. Entomol. Res. 67:* 659–666.

Fowler, H. G. & Robinson, S. W., 1979. Foraging by *Atta sexdens* (Formicidae, Attini): seasonal patterns, caste and efficiency. *Ecol. Entomol. 4:* 239–248.

Fowler, H. G. & Stiles, E. W., 1980. Conservative resource management by leaf-cutting ants: the role of foraging territories, an trails and environmental patchiness. *Sociobiology 5:* 25–42.

Fox, M. D., 1978. Changes in the ant community of coastal heath following sand mining. *Bull. Ecol. Soc. Aust. 8:* 9.

Fox, M. D. & Fox, B. J., 1982. Evidence for interspecific competition influencing ant species diversity in a regenerating heathland. this volume.

Freeland, W. J. & Janzen, D. H., 1974. Strategies in herbivory by mammals: the role of secondary compounds. *Am. Nat. 108:* 269–289.

Freese, C. H., 1976. Predation on swollen-thorn acacia ants by white-faced monkeys, *Cebus capucinus. Biotropica, 8:* 278–281.

Galle, L., 1972. Study of ant populations in various grassland ecosystems. *Acta biologica, Szeged, 18:* 159–164.

Galle, L., 1978. Dispersion of the nests of an ant species, *Lasius alienus* (Hymenoptera, Formicidae). *Acta Univ. Szeged. Acta Biol., 24:* 105–110.

Gamboa, G. J., 1975. Foraging and leaf cutting of the desert gardening ant, *Acromyrmex versicolor versicolor* (Hymenoptera, Formicidae). *Oecologia 20:* 103–110.

Garling, L., 1979. Origin of ant-fungus mutualism: a new hypothesis. *Biotropica 11:* 284–291.

Gates, B. N., 1940. Dissemination by ants of the seeds of *Trillium grandiflorum. Rhodora 42:* 194–196.

Gates, B. N., 1942. The dissemination by ants of the seeds of bloodroot, *Sanguinaria canadensis. Rhodora 44:* 13–15.

Gates, B. N., 1943. Carunculate seed dissemination by ants. *Rhodora 45:* 438–445.

Gentry, J. B. & Dunbar, J. M., 1977. Foraging territories of the Florida harvester ant, *Pogonomyrmex badius* (Hymenoptera, Formicidae). *Assoc. Southeast Biol. Bull. 24:* 53.

Gentry, J. B. & Stiritz, K. L., 1972. The role of the Florida harvester ant *Pogonomyrmex badius* in old field mineral nutrient relationships. *Environ. Entomol. 1:* 39–41.

Gilbert, L. E., 1979. Development of theory in the analysis of insect–plant interactions. In: Horn, D. J. (ed.), *Analysis of Ecological Systems,* 117–154. Ohio State U.P., Columbus.

Gilbert, L. E. & Raven, P. H. (eds.), 1975. *Coevolution in Animals and Plants.* Univ. Texas Press, Austin.

Gilbert, L. E. & Smiley, J. T., 1978. Determinants of local diversity in phytophagous insects: host specialists in tropical environments. *Symp. R. Entomol. Soc. London, 9:* 89–104.

Girling, D. J., 1978. The distribution and biology of *Eldana saccharina,* (Lepidoptera, Pyralidae) and its relationship to other stem borers in Uganda. *Bull. Entomol. Res. 68:* 471–488.

Glancey, B. M., 1979. Damage to corn by the red imported fire ants. *J. Ga. Entomol. Soc. 14:* 198–201.

Golley, F. B. & Gentry, J. B., 1964. Bioenergetics of the southern harvester ant *Pogonomyrmex badius. Ecology 45:* 217–225.

Gomez, L., 1974. Biology of the potato-fern *Solanopteris brunei. Brenesia 4:* 37–61.

Gomez, L., 1977. The *Azteca* ants of *Solanopteris brunei. Am. Fern J. 67:* 31.

Goncalves, C. R., 1967. As formigas cortadeiras da Amazonia, dos generos 'Atta' Fabr. e 'Acromyrmex' Mayr (Hym., Formicidae). *Atas Simp. Biota Amazonica (Zool.) 5:* 181–202.

Gordon, S. A., 1980. Analysis of twelve Sonoran Desert seed species preferred by the desert harvester ant. *Madrono 27:* 68–78.

Gosswald, K., 1954. Uber die Wirtschaftlichkeit des Masseneinsatzes der Roten Waldameise. *Z. Angew. Zool., 145*–185.

Gosswald, K., 1958. Neue Erfahrungen uber Einwirkung der Roten Waldameise auf den Massenwechsel von Schadinsekten sowie einige methodische Verbesserungen bei ihrem praktischen Einsatz. *Proc. X Int. Congr. Entomol., Montreal 4:* 567–571.

Gotwald, W. H., 1974. Foraging behavior of *Anomma* driver ants in Ghana cocoa farms (Hymenoptera, Formicidae). *Bull. Inst. Fondam. Afr. Noire Ser. A: Sci. Nat. 36:* 705–713.

Gotwald, W. H., 1978. Trophic ecology and adaptation in tropical old-world ants of the subfamiliy Dorylinae (Hymenoptera, Formicidae). *Biotropica 10:* 161–169.

Gray, B., 1971. Notes on the field behavior of 2 ant species, *Myrmecia desertorum* and *Myrmecia dispar* (Hymenoptera, Formicidae). *Insectes Soc. 18:* 81-94.

Green, G. W. & Sullivan, C. R., 1950. Ants attacking larvae of the forest tent caterpillar, *Malacosoma disstria* Hbn. *Can. Ent. 82:* 194-195.

Greenslade, P. & Greenslade, P. J. M., 1971. The use of baits and preservatives in pitfall traps. *J. Aust. Entomol. Soc. 10:* 253-260.

Greenslade, P. J. M., 1970. Observations on the inland variety (v. *viridiaeneus* Viehmeyer) of the meat ant *Iridomyrmex pur 'ureus* (Frederick Smith) (Hymenoptera: Formicidae). *J. Aust. Entomol. Soc. 9:* 227-231.

Greenslade, P. J. M., 1971a. Phenology of three ant species in the Solomon Islands. *J. Aust. Entomol. Soc. 10:* 241-252.

Greenslade, P. J. M., 1971b. Interspecific competition and frequency changes among ants in Solomon Islands coconut plantations. *J. Appl. Ecol. 8:* 323-352.

Greenslade, P. J. M., 1972. Comparative ecology of four tropical ant species. *Insectes Soc. 19:* 195-212.

Greenslade, P. J. M., 1974. Some relations of the meat ant, *Iridomyrmex purpureus* (Hymenoptera: Formicidae) with soil in South Australia. *Soil B iol. Biochem. 6:* 7-14.

Greenslade, P. J. M., 1975. Survey of the ant fauna of Kunoth Paddock, Hamilton Downs Station, NT. *CSIRO Tech. Mem., 2/1975.*

Greenslade, P. J. M., 1976. The meat ant *Iridomyrmex purpureus* (Hymenoptera: Formicidae) as a dominant member of ant communities. *J. Aust. Ento. Soc. 15:* 237-240.

Greenslade, P. J. M., 1979. *A Guide to Ants of South Australia.* South Australian Museum Special Educ. Bull. Series, Adelaide, 44 pp.

Greenslade, P. J. M. & Greenslade, P., 1973. Ants of a site in arid southern Australia. *Proc. VII Congr., IUSSI, 7:* 145-149, London.

Greenslade, P. J. M. & Greenslade, P., 1977. Some effects of vegetation cover and disturbance on a tropical ant fauna. *Insectes Soc. 24:* 163-182.

Greenslade, P. J. M. & Mott, J. J., 1978. Ants (Hymenoptera, Formicidae) of native and sown pastures in the Katherine area, N. T., Australia. 2nd *Aust. Conf. Grassl. Invert. Ecol., Session 5.* Palmerston North, New Zealand.

Griffiths, D., 1980. Foraging costs and relative prey size. *Am. Nat., 743-752.*

Groom, P., 1893. On *Dischidia rafflesiana* (Wall.) *Ann. Bot. 7:* 223-242.

Guerrant, E. O. & Fiedler, P. L., 1981. Flower defences against nectar-pilferage by ants. Biotropica suppl., *Reproductive Botany,* 25-33.

Haber, W. A., Frankie, G. W., Baker, H. G., Baker, I. & Koptur, S., 1982. Ants like flower nectar. *Biotropica,* in press.

Hagemann, W., 1969. Zur Morphologie der Knolle von *Polypodium bifrons* Hook. und *P. brunei* Werkle. *Soc. Bot. de France, Memoires,* 17-27.

Haines, B., 1971. *Plant Responses to mineral Nutrient Accumulations in Refuse Dumps of a Leaf-cutting Ant in Panama.* Dissertation, Duke University, N.C.

Haines, B., 1975. Impact of leaf-cutting ants on vegetation-development at Barro Colorado Island. In: Golley, F. B. & Medina, E. (eds.), *Tropical Ecological Systems,* 99-111. Springer, New York.

Haines, B. L., 1978. Element and energy flows through colonies of the leaf-cutting ant *Atta colombica* in Panama. *Biotropica 10:* 270-277.

Haines, I. H. & Haines, J. B., 1978a. Colony structure, seasonality and food requirements of the crazy ant *Anoplolepis longipes* in the Seychelles. *Ecol. Entomol. 3:* 109-118.

Haines, I. H. & Haines, J. B., 1978b. Pest status of the crazy ant *Anoplolepis longipes* (Hymenoptera, Formicidae) in the Seychelles. *Bull. Entomol. Res. 68:* 627-638.

Hall, B. M., 1762. *Dissertatio Botanica sistens Nectaria florum.* Upsala: *Dissertationes Academica* (Linne), *122.*

Hamilton, W. D. & May, R. M., 1977. Dispersal in stable habitats. *Nature 269:* 578-581.

Handel, S. N., 1976. Dispersal ecology of *Carex pedunculata* (Cyperaceae), a new North American myrmecochore. *Am. J. Bot. 63:* 1971-1979.

Handel, S. N., 1978a. The competitive relationship of three woodland sedges and its bearing on the evolution of ant dispersal of *Carex pedunculata. Evolution 32:* 151-163.

Handel, S. N., 1978b. New ant-dispersed species in the genera *Carex, Luzula* and *Claytonia. Can. J. Bot., 56:* 2925-2927.

Hansen, S. R., 1978. Resource utilization and coexistence of three species of *Pogonomyrmex* ants in an upper Sonoran grassland community. *Oecologia 35:* 109-117.

Harborne, J. B., 6ed.), 1978. *Biochemical Aspects of Plant and Animal Coevolution. Ann. Proc. Phytochem. Soc. Europe, 15.* Academic, London.

Hardy, A. D., 1912. The distribution of leaf glands in some Victorian Acacias. *Vic. Nat., 29:* 26-32.

Harkness, M. L. R. & Harkness, R. D., 1976. Functional differences between individual ants *Cataglyphis bicolor. J. Physiol. 25:* 124-125.

Harper, J. L., 1977. *Population Biology of Plants.* Academic Press, London.

Harper, J. L., Lovell, P. H. & Moore, K. G., 1970. The shapes and sizes of seeds. *Ann. Rev. Ecol. Syst. 1:* 327-356.

Harris, L. D., 1969. A consideration of the nutrient 'sumping' activities of leaf-cutter ants (*Atta* sp.) in the new world tropics. Research report in *Advanced Population Biology.* San Jose, Costa Rica: Organization for Tropical Studies.

Harris, P., 1973. Insects in the population dynamics of plants. In: van Emden, H. F. (ed.), *Insect/Plant Relationships.* Wiley, New York.

Harvey, D. J. & Webb, T. A., 1980. Ants associated with *Harkenclenus titus, Glaucopsyche lygdamus* and *Celastrina argiolus* (Lycaenidae). *J. Lepid. Soc. 34:* 371-372.

Haskins, C. P. & Haskins, E. F., 1965. *Pheidole megacephala* and *Iridomyrmex humilis* in Bermuda – equilibrium or slow replacement? *Ecology 46:* 736-740.

Hassell, M. P. & Southwood, T. R. E., 1978. Foraging strategies of insects. *Ann. Rev. Ecol. Syst. 9:* 75-98.

Hatch, M. H., 1933. Studies on the Leptodiridae (Catopidae) with descriptions of new species. *J. NY Entomol. Soc. 41:* 187-239.

Hayashida, K., 1960. Studies on the ecological distribution of ants in Sapporo and its vicinity (1 et 2), *Insectes Soc. 7:* 125-162.

Heim, D. R., 1898. The biologic relations between plants and ants.

Smithsonian Report for 1896, 411-455. Government Printing Office, Washington.

Heithaus, E. R., 1981. Seed predation by rodents on three ant-dispersed plants. *Ecology 62:* 136-145.

Heithaus, E. R., Culver, D. L. & Beattie, A. J., 1980. Models of some ant-plant mutualisms. *Am. Nat. 116:* 347-361.

Hemmingsen, A. M., 1977. Studies on worm-lions. *Entomol. Medd. 45:* 167-188.

Herbers, J. M., 1980. On caste ratios in ant colonies: population responses to changing environments. *Evolution 34:* 575-585.

Hervey, A. & Nair, M. S., 1979. Antibiotic metabolite of a fungus cultivated by gardening ants. *Mycologia 71:* 1064-1066.

Hervey, A., Rogerson, C. T. & Leong, I., 1977. Studies on fungi cultivated by ants. *Brittonia 29:* 226-236.

Hickman, J., 1974. Pollination by ants: a low-energy system. *Science 184:* 1290-1292.

Higashi, S. & Yamayuchi, K., 1979. Influence of a super colonial ant, *Formica yessensis*, on the distribution of other ants in Ishikari Coast, Japan. *Jpn. J. Ecol. 29:* 257-264.

Hill, M. G. & Blackmore, P. J. M., 1980. Interactions between ants and the coccid *Icerya seychellarum* on Aldabra Atoll. *Oecologia 45:* 360-365.

Hinton, H. E., 1951. Myrmecophilous Lycaenidae and other Lepidoptera. A summary. *Trans. So. Lond. Entomol. Nat. Hist. Soc. 1949-1950,* 111-174.

Hocking, B., 1970. Insect interactions with the swollen thorn acacias. *Trans. Roy. Entomol. Soc., London 122:* 211-255.

Hocking, B., 1975. Ant-plant mutalism: evolution and energy. In: Gilbert, L. E. & Raven, P. H. (eds.), *Coevolution of Animals and Plants*, 78-90. Univ. Texas Press, Austin.

Hodgson, E. S., 1955. An ecological study of the behaviour of the leaf-cutting ant *Atta cephalotes*. *Ecology 36:* 293-304.

Holldobbler, B., 1970. *Steatoda fulva* (Theridiidae): a spider that feeds on harvester ants. *Psyche 77:* 202-208.

Holldobbler, B., 1971. Homing in the harvester ant *Pogonomyrmex badius*. *Science 171:* 1149-1151.

Holldobbler, B., 1974. Home range orientation and territoriality in harvesting ants. *Proc. Natl. Acad. Sci. 71:* 3271-3277.

Holldobbler, B., 1976. Recruitment behavior, home range orientation and territoriality in harvester ants, *Pogonomyrmex. Behav. Ecol. Sociobiol. 1:* 3-44.

Holldobbler, B., 1980. Canopy orientation: a new kind of orientation in ants. *Science 210:* 86-88.

Holldobbler, B. & Lumsden, C. J., 1980. Territorial strategies in ants *Science 210:* 732-739.

Holldobbler, B., Stanton, R. C. & Markl, H., 1978. Recruitment and food retrieving behavior in *Novomessor* (Formicidae, Hymenoptera). I. Chemical signals. *Behav. Ecol. Sociobiol. 4:* 163-181.

Holldobbler, B. & Wilson, E. O., 1970. Recruitment trails in the harvester ant *Pogonomyrmex badius*. *Psyche 77:* 385-399.

Holt, S. J., 1955. On the foraging activity of the wood ant. *J. Anim. Ecol. 24:* 1-34.

Holttum, R. E., 1954. *Plant Life in Malaya*. Longmans, Green and Co., London.

Hopper, S., 1980. Pollination of the rain forest tree *Syzygium tierneyanum* (Myrtaceae) at Kuranda, Northern Queensland, Australia. *Aust. J. Bot. 28:* 223-238.

Horn, H. S., 1974. The ecology of secondary succession. *Ann.*

Rev. Ecol. Syst. 5: 25-37.

Horstmann, K., 1974. The foraging period of red ant *Formica polyctena* workers outside the nest. *Waldhygiene 10:* 241-246.

Horstmann, K., 1975. The behavior of foraging wood ants *Formica polyctena* in the spring. *Waldhygiene 11:* 1-12.

Horstmann, K., 1977. The structure of the nest of the forest ant and its significance to food transport (*Formica polyctena*). *Mitt. Dtsch. entomol. Ges. 35:* 91-98.

Horvitz, C. C. & Beattie, A. J., 1980. Ant dispersal of *Calathea* (Marantaceae) seeds by carnivorous ponerines (Formicidae) in a tropical rain forest. *Am. J. Bot. 67:* 321-326.

Hough, W. S., 1922. Observations on two mealybugs, *Trionymus trifolii* Forbes, and *Pseudococcus meritimus* Ehrh. *Entomol. News 33:* 171-176.

Howard, F. W. & Oliver, A. D., 1978. Arthropod populations in permanent pastures treated and untreated with mirex for red imported fire ant control. *Environ. Entomol. 7:* 901-903.

Hubbell, S. P., 1979. Tree dispersion, abundance and diversity in a tropical dry forest. *Science 203:* 1299-1309.

Hubbell, S. P., 1980. Seed predation and the coexistence of tree species in tropical forests. *Oikos 35:* 214-229.

Hubbell, S. P., Johnson, L. K., Stanislav, E., Wilson, B. & Fowler, H., 1980. Foraging by bucket-brigade in leaf-cutter ants. *Biotropica 12:* 210-213.

Hunt, G. L., 1977. Low preferred foraging temperatures and nocturnal foraging in a desert harvester ant. *Am. Nat. 111:* 589-591.

Hunt, J. H., 1974. Temporal activity patterns in 2 competing ant species (Hymenoptera, Formicidae). *Psyche J. Entomol. 81:* 237-242.

Huxley, C. R., 1978. The ant-plants *Myrmecodia* and *Hydnophytum* (Rubiaceae) and the relationships between their morphology, ant occupants, physiology and ecology. *New Phytol. 80:* 231-268.

Huxley, C. R., 1980. Symbiosis between ants and epiphytes. *Biol. Rev. 55:* 321-340.

Huxley, C. R., 1982. Ant-epiphytes of Australia, this volume.

Inouye, R. S., Byers, G. S. & Brown, J. H., 1980. Effects of predation and competition on survivorship, fecundity, and community structure of desert annuals. *Ecology 61:* 1344-1351.

Inouye, D. W. & Inouye, R. S., 1980. The amino acids of extrafloral nectar from *Helianthella quinquenervis* (Asteraceae). *Am. J. Bot., 67:* 1394-1396.

Inouye, D. W. & Taylor, O. R., 1975. A mutualistic ant-plant relationship from a high-altitude temperate region. *Am. Zool., 15:* 784.

Inouye, D. W. & Taylor, O. R., 1979. A temperate region plant-ant-seed predator system: consequences of extra floral nectar secretion by *Helianthella quinquenervis*. *Ecology, 60:* 1-7.

Iwanami, Y., Iwamatsu, M., Okada, I. & Iwadare, T., 1980. Comparison of inhibitory effects of royal jelly acid and myrmicacin on germination of *Camellia sinensis* pollens. *Experientia, 35:* 1311-1312.

Jaisson, P., 1980. Environmental preference induced experimentally in ants. (Hymenoptera, Formicidae). *Nature, 286,* 388-389.

Jakubczyk, H., Czerwinski, Z. & Petal, J., 1972. Ants as agents of the soil habitat changes. *Ecologia polska, 20:* 153–161.

Janda, C., 1931. Die extranuptialen Nektarien der Malvaceen. *Oesterr. Bot. Z., 86:* 81–130.

Jander, R. & Jander, U., 1979. An exact field test for the fade-out time of the odor trails of Asian weaver ants *Oecophylla smaragdina. Insectes Soc., 26:* 165–169.

Janzen, D. H., 1965. *The Interaction of the Bull's-Horn Acacia (Acacia cornigera* L.) *With One of its Ant Inhabitants (Pseudomyrmex fulvescens* Emery) *in Eastern Mexico.* Ph.D. Thesis, Univ. California, Berkeley.

Janzen, D. H., 1966. Coevolution of mutualism between ants and acacias in Central America. *Evolution 20:* 249–275.

Janzen, D. H., 1967a. Interaction of the Bull's-Horn acacia (*Acacia cornigera* L.) with an ant inhabitant (*Pseudomyrmex ferruginea* F. Smith) in eastern Mexico. *Kans. Univ. Sci. Bull. 47:* 315–558.

Janzen, D. H., 1967b. Fire, vegetation structure, and the ant x acacia interaction in Central America. *Ecology 48:* 26–35.

Janzen, D. H., 1969a. Allelopathy by myrmecophytes: the ant *Azteca* as an allelopathic agent of *Cecropia, Ecology 50:* 147–153.

Janzen, D. H., 1969b. Seed-eaters versus seed size, number, toxicity and dispersal. *Evolution 23:* 1–27.

Janzen, D. H., 1969c. Birds and the ant x acacia interactions in Central America, with notes on birds and other myrmecophytes. *Condor 71:* 240–256.

Janzen, D. H., 1970. Herbivores and the number of tree species in tropical forests. *Am. Nat. 104:* 501–528.

Janzen, D. H., 1971. Seed predation by animals. *Ann. Rev. Ecol. Syst. 2:* 465–492.

Janzen, D. H., 1972. Protection of *Barteria* (Passifloraceae) by *Pachysima* ants (Pseudomyrmecinae) in a Nigerian rain forest. *Ecology 53:* 885–892.

Janzen, D. H., 1973a. Dissolution of mutualism between *Cecropia* and its *Azteca* ants. *Biotropica 5:* 15–28.

Janzen, D. H., 1973b. Evolution of polygnous obligate acacia – ants in western Mexico. *J. Anim. Ecol. 42:* 727–750.

Janzen, D. H., 1974a. Tropical blackwater rivers, animals, and mast fruiting by the Dipterocarpaceae. *Biotropica 6:* 69–103.

Janzen, D. H., 1974b. Epiphytic myrmecophytes in Sarawak, Indonesia: mutualism through the feeding of plants by ants. *Biotropica 6:* 237–259.

Janzen, D. H., 1974c. Swollen-thorn acacias of Central America. *Smithson. Contr. Bot., 13.*

Janzen, D. H., 1977. Why don't ants visit flowers? *Biotropica 9:* 252.

Janzen, D. H., 1980. When is it coevolution? *Evolution 34:* 611–612.

Janzen, D. H. & McKey, D., 1977. *Musanga cecropiodes* is a *Cecropia* without its ants. *Biotropica 9:* 57.

Jeanne, R. L., 1979. A latitudinal gradient in rates of ant predation. *Ecology 60:* 1211–1224.

Jeffery, D. C., Arditti, J. & Koopowitz, H., 1970. Sugar content in floral and extrafloral exudates of orchids: pollination, myrmecology and chemotaxonomic implications. *New Phytol. 69:* 187–195.

Jensen, T. F., 1978. An energy budget for a field population of *Formica pratensis* (Hymenoptera, Formicidae). *Nat. Jutlandica 20:* 203–226.

Jermy, A. C. & Walker, T. G., 1975. *Lecanopteris spinosa* – a new ant-fern from Indonesia. *Fern Gazette 11:* 165–176.

Johns, G. G. & Greenup, L. R., 1976a. Pasture seed theft by ants in northern New South Wales. *Aust. J. Exp. Agric. Anim. Husb. 16:* 249.

Johns, G. G. & Greenup, L. R., 1976b. Predictions of likely theft by ants of oversown seed for the northern tablelands of New South Wales. *Aust. J. Exp. Agric. Anim. Husb. 16:* 257–264.

Jolivet, P., 1973. The myrmecophile plants of southeast Asia. *Cah. Pac., 17:* 41–69.

Jones, D. & Sterling, W. L., 1979. Manipulation of red imported fire ants, *Solenopsis invicta,* in a trap crop for boll weevil, *Anthonomus grandis,* suppression. *Environ. Entomol., 8:* 1073–1077.

Jonkman, J. C., 1978. Nests of the leaf-cutting ant, *Atta vollenweideri,* as accelerators of succession in pastures. *Z. Angew. Entomol., 86:* 25–34.

Jonkman, J. C., 1979. Population dynamics of leaf-cutting ant nests in a Paraguayan pasture. *Z. Angew. Entomol., 87:* 281–293.

Jutsum, A. R., Cherrett, J. M. & Fisher, M., 1981. Interactions between the fauna of citrus trees in Trinidad and the ants *Atta cephalotes* and *Azteca* sp. *J. Appl. Ecol. 18:* 187–196.

Jutsum, A. R., Saunders, T. S. & Cherrett, J. M., 1979. Intraspecific aggression in the leaf cutting ant *Acromyrmex octospinosus. Anim. Behav., 27:* 839–844.

Kajak, A., Breymeyer, A., Petal, J. & Olechowicz, E., 1972. The influence of ants of the meadow invertebrates. *Ekologia polska, 20:* 163–171.

Karsten, G., 1895. Morphologische und biologische Untersuchungen uber einige Epiphytenformen der Molukken. *Ann. Jard. Bot. Buitenz., 12:* 185–195.

Keeler, K. H., 1977. The extrafloral nectaries of *Ipomoea carnea* (Convolvulaceae). *Am. J. Bot. 64:* 1182–1188.

Keeler, K. H., 1979a. Distribution of plants with extrafloral nectaries and ants at 2 elevations in Jamaica. *Biotropica 11:* 152–154.

Keeler, K. H., 1979b. Species with extrafloral nectaries in a temperate flora (Nebraska) *Prairie Nat. 11:* 33–38.

Keeler, K. H., 1979c. Morphology and distribution of petiolar nectaries in *Ipomoea* (Convolvulaceae). *Am. J. Bot. 66:* 946–952.

Keeler, K. H., 1979d. Distribution of plants with extrafloral nectaries in temperate ecosystems and their relation to ant abundance. *Bull. Ecol. Soc. Am., 60:* 116.

Keeler, K. H., 1980. The extrafloral nectaries of *Ipomoea leptophylla* (Convolvulaceae). *Am. J. Bot., 67:* 216–222.

Keeler, K. H., 1980. Distribution of plants with extrafloral nectaries in temperate communities. *Am. Midl. Nat., 104:* 274–280.

Keeler, K. H., 1981a. Function of *Mentzelia nuda* (Loasaceae) postfloral nectaries in seed defense. *Am. J. Bot., 68:* 295–299.

Keeler, K. H., 1981b. Infidelity by acacia-ants. *Biotropica, 13:* 79–80.

Keeler, K. H., ms. Cover of plants with extrafloral nectaries at four Northern Californian sites.

Kendeigh, S. C. & West, G. C., 1965. Caloric values of plant seeds eaten by birds. *Ecology 46:* 553–555.

Kerner von Marilaun, A., 1978. *Flowers and their Unbidden*

Guests. (Transl. W. Ogle). Kegan Paul, London.

Kerr, A. F. G., 1912. Notes on *Dischidia rafflesiana* Wall., and *Dischidia nummularia* Br. *Scient. proc. R. Dubl. Soc., 13:* 293–309.

Kilham, L., 1979. Chestnut-coloured woodpeckers, *Celeus castaneus,* feeding as a pair on ants. *Wilson. Bull., 91:* 149–150.

King, T. J., 1976. The viable seed contents of ant-hill and pasture soil. *New Phytol. 77:* 143–147.

King, T. J., 1977a. The plant ecology of ant hills in calcareous grasslands. I. Patterns of species in relation to ant-hills in southern England. *J. Ecol. 65:* 235–256.

King, T. J., 1977b. The plant ecology of ant hills in calcareous grasslands. II. Succession on the mounds. *J. Ecol. 65:* 257–278.

King, T. J., 1977c. The plant ecology of ant hills in calcareous grasslands. III. Factors affecting the population sizes of selected species. *J. Ecol. 65:* 279–315.

King, T. J., 1981. Ant hills and grassland history. *J. Biogeog. 8:* 329–334.

Kitching, R. L., 1981. Egg clustering and the southern hemisphere lycaenids: comments on a paper by N. E. Stamp. *Am. Nat., 118:* 423–425.

Kitching, R. L. & Filshie, B. K., 1974. The morphology and mode of action of the anal apparatus of membracid nymphs with special reference to *Sextius virescens* (Fairmaire) (Homoptera). *J. Ent., (A), 49:* 81–88.

Kitching, R. L., Edwards, E. D., Ferguson, D., Fletcher, M. B. & Walker, J. M., 1978. The butterflies of the Australian Capital Territory. *J. Aust. Ent. Soc., 17:* 125–133.

Kleinfeldt, S. E., 1978. Ant-gardens: the interaction of *Codonanthe crassifolia* (Gesneriaceae) and *Crematogaster longispina* (Formicidae). *Ecology, 59:* 449–456.

Koptur, S., 1979. Facultative mutualism between weedy vetches bearing extrafloral nectaries and weedy ants in California, U.S.A. *Am. J. Bot., 66:* 1016–1020.

Laine, K. J. & Niemela, P., 1981. The influence of ants on the survival of mountain birches during an *Oporinia autumnata* (Lep., Geometridae) outbreak. *Oecologia 47:* 39–42.

Lamborn, W. A., 1913. On the relationship between certain West African insects, especially ants, Lycaenidae and Homoptera. *Trans. Ent. Soc. Lond., 61:* 436–512.

Ledoux, A., 1967. Action de la temperature sur l'activite d'*Aphenogaster senilis (testaceo-pilosa)* Mayr. (Hymenoptera, Formicoidea). *Insectes. soc. 14:* 131–156.

Leslie, J. K., 1965. Factors responsible for failures in the establishment of summer grasses on the black earths of the Darling Downs, Queensland. *Queensl. J. Agric. Anim. Sci., 22:* 17–38.

Levieux, J. & Diomande, T., 1978a. The nutrition of granivorous ants. Part 1. Cycle of activity and diet of *Messor galla* and *Messor regalis* (Hymenoptera, Formicidae). *Insectes Soc., 25:* 127–140.

Levieux, J. & Diomande, T., 1978b. The nutrition of granivorous ants. Part 2. Cycle of activity and diet of *Brachyponera senaarensis* (Hymenoptera, Formicidae). *Insectes Soc., 25:* 187–196.

Levieux, J. & Louis, D., 1975. Food of tropical ants. Part 2. Feeding behavior and diet of *Camponotus vividus* (Hymenoptera, Formicidae): intra generic comparisons. *Insectes Soc., 22:* 391–404.

Levin, D. A., 1971. Plant phenolics: and ecological perspective. *Am. Nat., 105:* 157–81.

Levin, D. A., 1973. The role of trichomes in plant defense. *Q. Rev. Biol., 48:* 3–15.

Levin, D. A., 1971. Plant phenolics: an ecological perspective. tive. *Am. Nat. 108:* 193–206.

Levins, R., Pressick, M. L. & Heatwole, H., 1973. Coexistence patterns in insular ants. *Am. Sci., 61:* 463–472.

Lewis, T., Pollard, G. V. & Dibley, G. C., 1974a. Rhythmic foraging in the leaf-cutting ant *Atta cephalotes* (Formicidae, Attini). *J. Anim. Ecol., 43:* 129–141.

Lewis, T., Pollard, G. V. & Dibley, G. C., 1974b. Micro-environmental factors affecting diel patterns of foraging in the leaf-cutting ant *Atta cephalotes* (Formicidae, Attini). *J. Anim. Ecol., 43:* 143–153.

Lidicker, W. Z., 1979. A clarification of interactions in ecological systems. *Bioscience, 29:* 475–477.

Lieberburg, I., Kranz, P. M. & Siep, A., 1975. Bermudan ants revisited: the status and interaction of *Pheidole megacephala* and *Iridomyrmex humilis. Ecology. 56:* 473–478.

Lind, N. K., 1981. Lytic mechanism of antimicrobial action of fire ant venoms. *Fed. Proc. 40:* 695.

Littledyke, M. & Cherrett, J. M., 1975. Variability in the selection of substrate by the leaf-cutting ants *Atta cephalotes* and *Acromyrmex octospinosus* (Formicidae, Attini). *Bull. Entomol. Res., 65:* 33–48.

Littledyke, M. & Cherrett, J. M., 1978a. Defense mechanisms in young and old leaves against cutting by the leaf-cutting ants *Atta cephalotes* and *Acromyrmex octospinosus* (Hymenoptera, Formicidae). *Bull. Entomol. Res., 68:* 263–272.

Littledyke, M. & Cherrett, J. M., 1978b. Olfactory responses of the leaf-cutting ants *Atta cephalotes* and *Acromyrmex octospinosus* (Hymenoptera, Formicidae) in the laboratory. *Bull. Entomol. Res., 68:* 273–282.

Lockwood, L. L., 1973. Distribution, density and dispersion of two species of *Atta* (Hymenoptera, Formicidae) in Guanacaste Province, Costa Rica. *J. Anim. Ecol., 42:* 803–818.

Lofgren, C. S., Banks, W. A. & Glancey, B. M., 1975. Biology and control of imported fire ants. *Ann. Rev. Entomol., 20:* 1–30.

Low, W. A. & Terrill, B., 1974. Density of mulga ants (*Rhytidoponera, Polyrachis:* Formicidae) in a mulga community in central Australia. *Grassl. Invert. Conf.,* Armidale.

Lu, K. L. & Mesler, M. R., 1981. Ant dispersal of a neotropical forest floor gesneriad. *Biotropica, 13:* 159–160.

Ludwig, J. A. & Whitford, W. G., 1979. Short-term water and energy flow in arid ecosystems. In: Noy-Meir, I. (ed.), *Ecosystem Dynamics.* Cambridge Univ. Press, London.

Lugo, A. E., Farnworth, E. G., Pool, D., Jerez, P. & Kaufman, G., 1973. The impact of the leaf-cutter ant *Atta colombica* on the energy flow of a tropical wet forest. *Ecology, 54:* 1292–1301.

Lyford, W. H., 1963. Importance of ants to brown podzolic soil genesis in New England. *Harvard Forest Paper, 7.* Harvard Univ. press, Cambridge, Mass.

Lynch, J. F., Balinsky, E. C. & Vail, S. C., 1980. Foraging patterns in three sympatric forest ant species, *Prenolepis imparis, Paratrechina melanderi* and *Aphaenogaster rudis* (Hymenoptera, Formicidae). *Ecol. Entomol., 5:* 353–372.

MacArthur, R. H., 1972. *Geographical Ecology.* Harper & Row, New York.

MacArthur, R. H. & Pianka, E. R., 1966. On the optimal use of a patchy habitat. *Am. Nat., 100:* 603–609.

Macedo, M. & Prance, G. T., 1978. Notes on the vegetation of Amazonia. II. The dispersal of plants in Amazonian white sand campinas: the campinas as functional islands. *Brittonia, 30:* 203–215.

Mahdihassan, S., 1978. The mango tree in Karachi, Pakistan, as extensively infected by a scale insect. *Pak. J. Sci. Ind. Res., 21:* 19–21.

Majer, J. D., 1972. The ant mosaic in Ghana cocoa farms. *Bull. Ent. Res. 62:* 151–160.

Majer, J. D., 1976. The influence of ants and ant manipulation on the cocoa farm fauna. *J. Appl. Ecol., 13:* 157–175.

Majer, J. D., 1978a. Preliminary survey of the epigaeic invertebrate fauna with particular reference to ants, in areas of different land use at Dwellingup, Western Australia. *Forest Ecol. Manage. 1:* 32–34.

Majer, J. D., 1978b. The seedy side of ants. *Gazette, 11:* 7–9.

Majer, J. D., 1978c. The importance of invertebrates in successful land reclamation with particular reference to bauxite mine rehabilitation. *Proc. Symp. Rehabilitation of Mined Lands in Western Australia* (Perth, 11 Oct 1978), 47–61. WAIT, Bentley, W.A.

Majer, J. D., 1978d. The influence of blanket and selective spraying on ant distribution in a West African cocoa farm. *Rev. Theobroma (Brasil), 8:* 87–93.

Majer, J. D., 1979. The possible protective function of extrafloral nectaries of *Acacia salicina.* Mulga Research Centre, *Annual Report, 2.* Perth.

Majer, J. D., 1980. The influence of ants on broadcast and naturally spread seeds in rehabilitated bauxite mined areas. *Reclamation Review 3:* 3–9.

Majer, J. D., 1982. Ant–plant interactions in the Darling Botanical District of Western Australia, this volume.

Majer, J. D., Day, J. E., Kabay, E. D. & Perriman, W. S., 1982. Recolonization by ants in bauxite mines rehabilitated by a number of different methods, in press.

Majer, J. D., Portlock, C. C. & Sochacki, S. J., 1979. Ant–seed interactions in the northern jarrah forest. *Abstr. Symp. Biol. Native Aust. Plants 25. Perth.*

Malicky, H., 1970. New aspects on the association between lycaenid larvae (Lycaenidae) and ants (Formicidae, Hymenoptera). *J. Lep. Soc., 24:* 190–202.

Mann, W. M., 1912. Parabiosis in Brazilian ants. *Psyche, 19:* 36–41.

Mann, W. M., 1919. The ants of the British Solomon Islands. *Bull. Mus. Comp. Zool. Harv. 63:* 362.

Mann, W. M., 1921. The ants of the Fiji Islands. *Bull. Mus. Comp. Zool. Harv. 64:* 401–499.

Mares, M. A. & Rosenzeig, M. L., 1978. Granivory in North and South American deserts: rodents, birds and ants. *Ecology, 59:* 235–241.

Markin, G. P., 1970. Foraging behavior of the Argentine ant in a California citrus grove. *J. Ecol. Entomol., 63:* 740–744.

Marking, G. P., O'Neal, J. & Collins, H. L., 1974. Effects of Mirex on the general ant fauna of a treated area in Louisiana. *Environ. Entomol., 3:* 895–897.

Markin, G. P., O'Neal, J. & Dillier, J., 1975. Foraging tunnels of the red imported fire ant *Solenopsis invicta* (Hymenoptera, Formicidae). *J. Kans. Entomol Soc., 48:* 83–89.

Marshall, D. L., Beattie, A. J. & Bollenbacher, W. E., 1979. Evidence for diglycerides as attractants in an ant (*Aphaenogaster rudis*) – seed interaction. *J. Chem. Ecol. 5:* 335–344.

Marshall, J. J. & Rickson, F. R., 1973. Characterization of the a-D-glucan from the plastids of *Cecropia peltata* as a glycogen-type polysaccharide. *Carbohydr. Res., 28:* 31–37.

Martin, M. M., 1970. The biochemical basis of the fungus-attine symbiosis. *Science, 169:* 16–20.

Martin, M. M., 1974. Biochemical ecology of the attine ants (Hymenoptera, Formicidae). *Acc. Chem. Res., 7:* 1–5.

Martin, M. M., Boyd, N. D., Gieselmann, M. J. & Silver, R. G., 1975. Activity of faecal fluid of a leaf-cutting ant toward plant cell wall polysaccharides. *J. Insect Physiol., 21:* 1887–1892.

Martin, M. M., Carls, C. A., Hutchins, R. F. N., MacConnell, J. G., Martin, J. S. & Steiner, O. D., 1967. Observations on *Atta colombica tonsipes* (Hymenoptera, Formicidae). *Ann. Ent. Soc. Am., 60:* 1329–1330.

Martin, M. M., Carman, R. M. & MacConnell, J. G., 1969. Nutrients derived from the fungus cultured by the fungus-growing ant *Atta colombica tonsipes. Ann. Ent. Soc. Am., 62:* 11–13.

Martin, M. M., Gieselman, M. J. & Martin, J. S., 1973. Rectal enzymes of attine ants: amylase and chitinase. *J. Insect Physiol., 19:* 1409–1416.

Martin, M. M., MacConnell, M. G. & Gale, G. R., 1969. The chemical basis for the attine ant-fungus symbiosis: absence of antibiotics. *Ann. Ent. Soc. Am., 62:* 386–388.

Martin, M. M. & Martin, J. S., 1970. The biochemical basis for the symbiosis between the ant, *Atta colombica tonsipes,* and its food fungus. *J. Insect Physiol., 16:* 109–119.

Martin, M. M. & Martin, J. S., 1971. The presence of protease activity in the rectal fluid of primitive attine ants. *J. Insect Physiol., 17:* 1897–1906.

Martin, M. M. & Weber, N. A., 1969. The cellulose utilizing capability of the fungus cultured by the attine ant *Atta colombica tonsipes. Ann. Ent. Soc. Am., 62:* 1386–1387.

Matsuda, K. & Sugawara, F., 1980. Defensive secretion of chrysomelid larvae, *Chrysomela vigintipunctata costella, Chrysomela populi* and *Gastrolina depressa* (Coleoptera, Chrysomelidae). *Appl. Entomol. Zool., 15:* 316–320.

McCook, H. C., 1879. *The Natural History of the Agricultural Ant of Texas. Acad. Nat. Sci.,* Philadelphia.

McCubbin, C., 1971. *Australian Butterflies,* Nelson, Melbourne.

McDade, L. A. & Kinsman, S., 1980. The impact of floral parasitism in 2 neotropical hummingbird pollinated plant species. *Evolution, 34:* 944–958.

McEvoy, P. B., 1979. Advantages and disadvantages to group living in treehoppers (Homoptera, Membracidae). *Misc. Publ. Entomol. Soc. Am., 11:* 1–14.

McGowan, A. A., 1969. Effect of seed-harvesting ants on the persistence of Wimmera rye-grass in pastures of northeast Victoria. *Aust. J. Exp. Agric. Anim. Husb. 9:* 37–40.

McKey, D., 1974. Ant-plants: selective eating of an unoccupied *Barteria* by a colobus monkey. *Biotropica, 6:* 269–270.

McKey, D., 1975. The ecology of coevolved seed dispersal systems. In: Gilbert, L. E. & Raven, P. (eds.), *Coevolution of Animals and Plants,* 159–191. Univ. Texas Press, Austin.

McKey, D., ms. Interaction of the ant-plant *Leonardoxa africana*

154

(Caesalpiniaceae) with its obligate inhabitants in rainforests in Cameroon.

McLain, D. K., 1980. Relationships among ants, aphids and coccinellids on wild lettuce, *Lactuca canadensis. J. Ga. Entomol. Soc., 15:* 417–418.

Meyen, F. J. F., 1837. *Ueber die Secretionsorgane der Pflanzen.* Berlin.

Miehe, H., 1911a. Untersuchungen uber die javanische *Myrmecodia.* In: Javanische Studien 2. *Abh. dt. mathematisch-physischen Klasse der Koniglich Sachsischen Ges. der Wissenschaften, 32:* 312–361.

Miehe, H., 1911b. Uber die javanische *Myrmecodia* und die Beziehung zu ihren Ameisen. *Biol. Zbl. 31:* 733–737.

Milewski, A. V. & Bond, W. J., 1982. Convergence of myrmecochory in mediterranean Australia and South Africa, this volume.

Mintzer, A., 1979. Foraging activity of the mexican leaf-cutting ant, *Atta mexicana,* in a Sonoran Desert habitat (Hymenoptera, Formicidae). *Insectes Soc., 26:* 364–372.

Mintzer, A., 1980. Simultaneous use of a foraging trail by 2 leaf-cutter ant species, *Acromyrmex versicolor* and *Atta mexicana,* in the Sonoran Desert, U.S.A. *J. N.Y. Entomol. Soc., 88:* 102–105.

Mobbs, C. J., Tedder, G., Wade, A. M. & Williams, R., 1978. A note on food and foraging in relation to temperature in the meat ant *Iridomyrmex purpureus viridianeus. J. Aust. Entomol. Soc., 17:* 193–198.

Montanucci, R. R., 1981. Habitat separation between *Phrynosoma douglassi* and *Phrynosoma orbiculare* (Lacetilia, Iguanidae) in Mexico. *Copeia:* 147–153.

Mordechai, J. B. & Kugler, K., 1976. Ecology of ants in the desert loess plain Sede-Zin of Sede Boger, Central Negev. *Isr. J. Zool., 25:* 216–217.

Morrill, W. L., 1978. Red imported fire ant predation on the alfalfa weevil and pea aphid. *J. Econ. Entomol., 71:* 867–868.

Morris, M. G., 1980. Ecology and evolution of rare British Papilionoidea (Lepidoptera). *Int. Cong. Syst. Evol. Biol., 2.* UBC, Vancouver, Canada.

Morton, S. R., 1979. Diversity of desert-dwelling mammals: a comparison of Australia and North America. *J. Mammal. 60:* 253–264.

Morton, S. R., 1982. Granivory in the Australian arid zone: diversity of harvester ants and structure of their communities. In: Barker, W. R. & Greenslade, P. J. M. (eds.), *Evolution of the Flora and Fauna of Arid Australia.* Peacock, Adelaide, in press.

Moser, J. C., 1963. Contents and structure of *Atta texana* nest in summer. *Ann. Ent. Soc. Am., 56* (3), 286–291.

Moser, J. C., 1967. Trails of the leafcutters. *Natural History, 76:* 32–35.

Moser, J. C. & Blum, M. S., 1963. Trailmarking substance of the Texas leaf-cutting ant: source and potency. *Science, 140:* 1228.

Mott, J. J. & McKeon, G. M., 1977. A note on the selection of seed types by harvester ants in northern Australia. *Aust. J. Ecol. 2:* 231–235.

Mudd, A. & Bateman, G. L., 1979. Rates of growth of the food fungus of the leaf-cutting ant *Atta cephalotes* (Hymenoptera, Formicidae) on different substrates gathered by the ants. *Bull. Entomol. Res., 69:* 141–148.

Mudd, A., Peregrine, D. J. & Cherrett, J. M., 1978. The chemical basis for the use of citrus pulp as fungus garden substrate by the leaf-cutting ants *Atta cephalotes* and *Acromyrmex octospinosus* (Hymenoptera, Formicidae). *Bull. Entomol Res., 68:* 673–686.

Mullenax, C. H., 1979. The use of jackbean (*Canavalia ensiformis*) as a biological control for leafcutting ants (*Atta* spp). *Biotropica, 11:* 313.

Muller, F., 1874. The habits of various insects. *Nature, 10:* 102–103.

Muller, F., 1880. Die Imbauba und Ihre Beschutzer. *Kosmos, 8:* 109–116.

Müller, P., 1933. Verbreitungsbiologie der Gariqueflora. *Beih. Bot. Zentralb. 50:* 395–469.

Nair, M. S. & Hervey, A., 1979. Structure of lepiochlorin, an antibiotic metabolite of a fungus (*Lepiota* sp.) cultivated by ants (*Cyphomyrmex costatus*). *Phytochemistry, 18:* 326–327.

Nelson, J. F. & Chew, R. M., 1977. Factors affecting the seed reserves in the soil of a Mojave Desert ecosystem, Rock Valley, Nye County, Nevada. *Am. Midl. Nat. 97:* 300–320.

Nesom, G. L., 1981. Ant dispersal in *Wedelia hispida* (Heliantheae, Compositae). *Southwest Nat. 26:* 5–12.

Neto, G. G. & Asakawa, N, M., 1978. Estudo de Mirmecodomaceos em algumas especies de Boraginaceae, Chrysobalanaceae, Melastomaceae e Rubiaceae. *Acta Amazonica, 8:* 45–50.

Newcomer, E. J., 1912. Some observations on the relations of ants and lycaenid caterpillars, and a description of the relational organs of the latter. *J. New York Ent. Soc., 20:* 31–36.

Nickle, D. A. & Neal, J. M., 1972. Observations on the foraging behavior of the southern harvester ant *Pogonomyrmex badius. Fla. Entomol., 55:* 65–66.

Nilsson, L. A., 1978. Pollination ecology of *Epipactis palustris* (Orchidaceae). *Bot. Not., 131:* 355–368.

Nixon, G. E. J., 1951. *The Association of Ants with Aphids and Coccids.* Commonwealth Institute of Entomology, London.

Norberg, R. A., 1977. An ecological theory on foraging time and energetics and choice of optimal food-searching method. *J. Anim. Ecol., 46:* 511–529.

Nordhagen, R., 1932a. Uber die Einrollung der Fruchtstiele bei der Gattung *Cyclamen* und ihre biologische Bedeutung. *Beih. Bot. Centralbl., 49:* 359–395.

Nordhagen, R., 1932b. Zur Morphologie und Verbreitungsbiologie der Gattung *Roscoea* SM. *Univ. Bergen Arbok. Naturvitensk., 4:* 1–57.

Nordhagen, R., 1959. Remarks on some new or little known myrmecochorous plants from North America and East Asia. *Bull. Res. Counc. Isr., Sect. D Bot., 7:* 184–201.

Noy-Meir, I., 1973. Desert ecosystems: environment and producers. *Ann. Rev. Ecol. Syst. 4:* 25–51.

O'Donoghue, J. G. & St. John, P. R. H., 1913. Further notes on the Brisbane Range, *Vict. Nat., 29:* 130–138.

O'Dowd, D. J., 1979. Foliar nectar production and ant activity on a neotropical tree, *Ochroma pyramidale. Oecologia, 43:* 233–248.

O'Dowd, D. J., 1980. Pearl bodies of a neotropical tree, *Ochroma pyramidale:* ecological implications. *Am. J. Bot., 67:* 543–549.

O'Dowd, D. J. & Hay, M. E., 1980. Mutualism between harvester ants and a desert ephemeral: seed escape from rodents. *Ecology 61:* 531–540.

Odum, E. P., 1971. *Fundamentals of Ecology.* Saunders, W. B., Philadelphia, Pennysylvania, USA.

Olive, J., 1978. Notes on the life history of *Jalmenus clementi* (Lepidoptera, Lycaenidae). *Aust. Entomol Mag., 4:* 115.

Orians, G. H. & Solbrig, O.T., 1977. *Convergent Evolution in Warm Deserts.* Dowden, Hutchinson and Ross, Stroudsburg, Pennsylvania.

Oster, G. F. & Wilson, E. O., 1978. *Caste and Ecology in the Social Insects. Monogr. Popul. Biol., 12.* Princeton Unive. Press, Princeton.

Owen, D. F., 1980. How plants may benefit from the animals that eat them. *Oikos, 35:* 230–235.

Paoli, G., 1929. Strane abitazione di una formica su acacie della Somalia. *Riv. Colonie Ital., 3:* 474–485.

Paoli, G., 1930. Contributo allo studio dei rapporti fra le acacie e le formiche. *Mem. Soc. Ent. Ital., 9:* 129–195.

Parker, G. H., 1925. The weight of vegetation transported by tropical fungus ants. *Psyche, 32:* 227–228.

Parsons, L. R., 1968. Aspects of leaf-cutter ant behaviour: *Acromyrmex octospinosa* and *Atta cephalotes.* Research Report in *Tropical biology: an Ecological Approach.* San Jose, Costa Rica: Organization for Tropical Studies.

Pavlova, Z. F., 1977. Earth hummocks inhabited by ants as the principal micro structures of lake coastal biogeocenoses. *Ekologiya, 5:* 62–71.

Penzig, O., 1892. Ueber die Perldrusen des Weinstockes und anderer Pflanzen. *Atti del Congr. Bot. Intern. (Genoa),* 237–245.

Peregrine, D. J. & Mudd, A., 1974. The effect of diet on the composition of the postpharyngeal glands of *Acromyrmex octospinosus* (Reich). *Insectes Soc., 21:* 417–424.

Peregrine, D. J., Mudd, A. & Cherrett, J. M., 1973. Anatomy and preliminary chemical analysis of the post-pharyngeal glands of the leaf-cutting ant *Acromyrmex octospinosus* (Reich) (Hymenoptera, Formicidae). *Insectes Soc., 20:* 355–363.

Peregrine, D. J., Percy, H. C. & Cherrett, J. M., 1972. Intake and possible transfer of lipid by the post-pharyngeal glands of *Atta cephalotes* (L.). *Entomologia experimentalis et applicata, 15:* 248–249.

Petal, J., 1978. The role of ants in ecosystems. In: Brian, M. V. (ed.), *Production Ecology of Ants and Termites,* 293–325. Cambridge UP, Cambridge.

Petal, J., 1980. Ant populations, their regulation and effect on soil in meadows. *Ekol. Pol. 28:* 297–326.

Petal, J., Jakubczyk, H. & Wojcik, Z., 1970. L'influence des fourmis sur la modification des sols et des plantes dans le milieu des praires. In: Phillipson, J. (ed.), *Methods of Study in Soil Ecology,* 235–240. UNESCO, Paris.

Petal, J., Nowak, E., Jakubczyk, H. & Czerwinski, Z., 1977. Effect of ants and earthworms on soil habitat modification. In: Lohm, V. & Persson, T. (eds.), *Soil Organisms as Components of Ecosystems,* 501–503. Swedish Nat. Sci. Res. Council, Stockholm.

Petersen, B., 1977. Pollination by ants in the alpine tundra of Colorado, USA. *Trans. Ill. State Acad. Sci., 70:* 349–355.

Pickett, C. H. & Clark, W. D., 1979. The function of extrafloral nectaries in *Opuntia acanthocarpa* (Cactaceae). *Am. J. Bot, 66:* 618–625.

Pierce, N. E. & Mead, P. S., 1981. Parasitoids as selective agents in the symbiosis between lycaenid butterfly larvae and ants. *Science, 211:* 1185–1187.

Pisarski, B., 1978. Comparison of various biomes. In: Brian, M. V. (ed.), *Production Ecology of Ants and Termites,* 326–331. Cambridge UP, Cambridge.

Pollard, G. V., 1973. *Some Factors Affecting the Foraging Behaviour in Leaf-Cutting Ants of the Genera Atta and Acromyrmex.* Ph.D. thesis, University of the West Indies, St. Augustine, Trinidad.

Pontin, A. J., 1958. A preliminary note on the eating of aphids by ants of the genus *Lasius* (Hymenoptera, Formicidae). *Entomol. Monthly Mag., 94:* 9–11.

Pontin, A. J., 1961. Population stabilization and competition between the ants *Lasius flavus* (F.) and *L. niger* (L.). *J. Anim. Ecol. 30:* 47–54.

Pontin, A. J., 1963. Further considerations of competition and the ecology of the ants *Lasius flavus* (F.) and *L. niger* (L.). *J. Anim. Ecol. 32:* 565–574.

Pontin, A. J., 1978. The numbers and distribution of subterranean aphids and their exploitation by the ant *Lasius flavus* (Fabr.). *Ecol. Ent. 3:* 203–207.

Porter, S. D. & Jorgensen, C. D., 1980. Recapture studies of the harvester ant, *Pogonomyrmex owhyeei* Cole, using a fluorescent marking technique. *Ecol. Entomol., 5:* 263–269.

Prance, G. T., 1973. Gesneriads in the ant gardens of the Amazon. *Gloxinian, 23:* 27–28.

Price, P. W., Bouton, C. E., Gross, P., McPheron, B. A., Thompson, J. N. & Weis, A. E., 1980. Interactions among three trophic levels: influence of plants on interactions between insect herbivores and natural enemies. *Ann. Rev. Ecol. Syst., 11:* 41–65.

Proctor, M. & Yeo, P., 1972. *The Pollination of Flowers.* Taplinger, Oxford.

Pudlo, R. J., Beattie, A. J. & Culver, D. C., 1980. Population consequences of changes in an ant-seed mutualism in *Sanguinaria canadensis. Oecologia, 146:* 32–37.

Pulliam, H. R., 1974. On the theory of optimized diets. *Am. Nat., 108:* 59–74.

Pulliam, H. R. & Brand, M. R., 1975. The production and utilization of seeds in plains grasslands of southeastern Arizona. *Ecology 56:* 1158–1166.

Quinlan, R. J. & Cherrett, J. M., 1978. Aspects of the symbiosis of the leaf-cutting ant *Acromyrmex octospinosus* and its food fungus. *Ecol. Entomol., 3:* 221–230.

Quinlan, R. J. & Cherrett, J. M., 1979. The role of fungus in the diet of the leafcutting ant *Atta cephalotes. Ecol. Entomol., 4:* 151–160.

Raciborski, M., 1898. Biologische mittheilungen aus Java. *Flora, 85:* 325–367.

Raciborski, M., 1900. Ueber myrmecophile Pflanzen. *Flora, 87:* 38–45.

Rauh, W., 1955. Botanische Mitteilungen aus den Anden. I. Morphologische und antomische Beobachtungen an *Polypodium bifrons* Hook. *Abh. der mathematisch-naturwissenschaftli-*

chen Klasse, Akademie der Wissenschaften und der Literatur, Mainz, 3: 3–15.

Rauh, W., 1973. *Solanopieris bismarkii* Rauh. *Trop. Subtrop. Pflanzenwelt, 5:* 223–256.

Ray, T. S. & Andrews, C. C., 1980. Ant-butterflies: butterflies that follow army ants, *Eciton burchelli*, to feed on antbird droppings. *Science, 210:* 1147–1148.

Rehr, S. S., Feeny, P. P. & Janzen, D. H., 1973. Chemical defenses in Central American non-ant *Acacias. J. Anim. Ecol. 42:* 405–416.

Reichman, O. J., 1974. *Some Ecological Factors of the Diets of Sonoran Desert Rodents.* Ph.D. Dissertation. Univ. N. Ariz.

Reichman, O. J., 1975. Relationships between dimensions, weights, volumes, and calories of some Sonoran Desert seeds. *Southwest Naturalist 20:* 573–575.

Reichman, O. J., 1976. Seed distribution and the effect of rodents on germination of desert annuals. US/IBP *Desert Biome Res. Mem. 76:* 20–26. Utah State Univ., Logan.

Reichman, O. J., 1979. Desert granivore foraging and its impact on seed densities and distributions. *Ecology 60:* 1085–1092.

Reichman, O. J. & Oberstein, O., 1977. Selection of seed distribution types by *Dipodomys merriami* and *Perognathus amplus. Ecology 58:* 636–643.

Rettig, E., 1904. Ameisenpflanzen-pflanzenameisen. *Beih. bot. Zbl., 17:* 89–121.

Reznikova, Z. H. & Kulikov, A. V., 1978a. Feeding and interaction of steppe ant species (Hymenoptera, Formicidae). *Entomol. Rev. 57:* 43–51.

Reznikova, Z. H. & Kulikov, A. V., 1978b. Feeding characteristics and interactions of different species of steppe ants (Hymenoptera, Formicidae). *Entomol Obozr. 57:* 68–80.

Rhoades, D. F., 1979. Evolution of plant chemical defense against herbivores. Pages 3–54. In Rosenthal, G. A. & Janzen, D. H. (eds.) *Herbivores: their Interaction with Secondary Plant Metabolites.* Academic Press, New York.

Rhoades, D. F. & Bergdahl, J. C., 1980. Adaptive significance of toxic nectar. *Am. Nat., 117:* 798–803.

Rhoades, D. F. & Cates, R. G., 1976. Toward a general theory of plant anti-herbivore chemistry. In: Wallace, J. W. & Mansell, R. L. (eds.), *Biochemical Interaction Between Plants and Insects. Recent Adv. Phytochem., 10:* 168–313. Plenum, New York.

Rice, B. L. & Westoby, M., 1981. Myrmecochory in sclerophyll vegetation of the West Head, N.S.W. *Aust. J. Ecol. 6:* 291–298.

Rickson, F. R., 1968. Nuclear and cytoplasmic tubules in cortical cells of leaf *Beltian* bodies. *J. Cell. Biol. 38:* 471–474.

Rickson, F. R., 1969. Developmental aspects of the shoot apex, leaf and Beltian bodies of *Acacia cornigera* L. *Am. J. Bot. 56:* 196–200.

Rickson, F. R., 1971. Glycogen plastids in Mullerian body cells of *Cecropia peltata*, a higher green plant. *Science, 173:* 344–347.

Rickson, F. R., 1973. Review of glycogen plastid differentiation in Mullerian body cells of *Cecropia peltata. Ann. N. Y. Acad. Sci. 210:* 104–114.

Rickson, F. R., 1975. The ultrastructure of *Acacia cornigera* L. Beltian body tissue. *Am. J. Bot. 62:* 913–922.

Rickson, F. R., 1976a. Anatomical development of the leaf trichilium and Mullerian bodies of *Cecropia peltata* L. *Am. J. Bot., 63:* 1266–1271.

Rickson, F. R., 1976b. Ultrastructural differentation of the Mullerian body glycogen plastid of *Cecropia peltata* L. *Am. J. Bot., 63:* 1272–1279.

Rickson, F. R., 1977. Progressive loss of ant-related traits of *Cecropia peltata* on selected Caribbean islands. *Am. J. Bot., 64:* 585–592.

Rickson, R. R., 1979a. Ultrastructural development of the beetle food tissue of *Calycanthus* flowers. *Am. J. Bot. 66:* 80–86.

Rickson, F. R., 1979b. Absorption of animal tissue breakdown products into a plant stem: the feeding of a plant by ants. *Am. J. Bot. 66:* 87–90.

Rickson, F. R., 1980. Developmental anatomy and ultrastructure of the ant food bodies (Beccarian bodies) of *Macaranga triloba* and *Macaranga hypoleuca* (Euphorbiaceae). *Am. J. Bot. 67:* 285–292.

Rico-Gray, V., 1980. Ants and tropical flowers. *Biotropica 12:* 223.

Ridley, H. N., 1910. Symbiosis of ants and plants. *Ann. Bot. 24:* 457–483.

Ridley, H. N., 1923. *The Flora of the Malay Peninsula.* Vol. II–*Gamopetalae.* Reeve, London.

Ridley, H. N., 1930. *The Dispersal of Plants Throughout the World.* Reeve, Ashford, Kent.

Risch, S., 1981. Ants as important predators of rootworm eggs in the neotropics. *J. Econ. Entomol. 74:* 88–90.

Risch, S., McClure, M., Vandemeer, J. & Waltz, S., 1977. Mutualism between three species of tropical *Piper* (Piperaceae) and their ant inhabitants. *Am. Midl. Nat. 98:* 433–443.

Risch, S. J. & Rickson, F. R., 1981. Mutualism in which ants must be present before plants produce food bodies. *Nature 291:* 149–150.

Rissing, S. W., 1981. Foraging specializations of individual seed-harvester ants. *Behav. Ecol. Sociobiol. 9:* 149–152.

Rissing, S. W. & Wheeler, J., 1976. Foraging responses of *Veromessor pergandei* to changes in seed production (Hymenoptera, Formicidae). *Pan-Pac Entomol. 52:* 63–72.

Robertson, C., 1897. Seed crests and myrmecophilous dissemination in certain plants. *Bot. Gaz. 23:* 288–289.

Rockwood, L. L., 1973a. The effect of defoliation on seed production in six Costa Rican tree species. *Ecology 54:* 1363–1369.

Rockwood, L. L., 1973b. Distribution, density and dispersion of two species of *Atta* (Hymenoptera: Formicidae) in Guanacaste Province, Costa Rica. *J. Anim. Ecol. 42:* 803–817.

Rockwood, L. L., 1975. The effects of seasonality on foraging in 2 species of leaf-cutting ants *Atta* in Guanacaste Province, Costa Rica. *Biotropica 7:* 176–193.

Rockwood, L. L., 1976. Plant selection and foraging patterns in two species of leaf cutting ants (*Atta*). *Ecology 57:* 48–61.

Rockwood, L. L., 1977. Foraging patterns and plant selection in Costa-Rican leaf cutting ants. *J. N. Y. Entomol. Soc. 85:* 222–233.

Rockwood, L. L. & Glander, K. E., 1979. Howling monkeys and leaf-cutting ants: comparative foraging in a tropical deciduous forest. *Biotropica 11:* 1–10.

Rogers, L. E, 1974. Foraging activity of the western harvester ant in the shortgrass plains ecosystem. *Environ. Entomol. 3:* 420–424.

Rogers, L. E. & Lavigne, R. J., 1974. Environmental effects of western harvester ants on the shortgrass plains ecosystem.

Environ. Entomol. 3: 994–997.

Rogers, L. E., Lavigne, R. J. & Miller, J. L., 1972. Bioenergetics of the western harvester ant in the short grass plains ecosystem. *Environ. Entomol. 1:* 763–768.

Room, P. M., 1975a. Diversity and organization of the ground foraging ant faunas of forest, grassland and tree crops in Papua New Guinea. *Aust. J. Ecol. 23:* 71–89.

Room, P. M., 1975b. Relative distributions of ant species in cocoa plantations in Papua New Guinea. *J. Appl. Ecol. 12:* 47–61.

Rosengren, R., 1971. Route fidelity, visual memory and recruitment behaviour in foraging wood ants of the genus *Formica. Acta Zool. Fennica 133:* 1–106.

Rosengren, R., 1977a. Foraging strategy of wood ants, *Formica rufa* group. Part 1. Age polyethism and topographic traditions. *Acta Zool. Fenn. 149:* 1–30.

Rosengren, R., 1977b. Foraging strategy of wood ants, *Formica rufa* group. Part 2. Nocturnal orientation and diel periodicity. *Acta. Zool. Fenn. 150:* 1–30.

Ross, G. N., 1966. Life-history studies on Mexican butterflies IV. The ecology and ethology of *Anatole rossi,* a myrmecophilous metalmark (Lepidoptera: Riodinidae). *Ann. Ent. Soc. Am. 59:* 985–1004.

Rouppert, C., 1926. Observations sur les perlules de diverses especes de phanerogames. *Mus. Hist. Nat. Bull., Paris 32:* 102–107.

Russell, M. J., Coaldrake, J. E. & Sanders, A. M., 1967. Comparative effectiveness of some insecticides, repellents and seed-pelleting devices in the prevention of ant removal of pasture seeds. *Trop. Grassl. 1:* 153–166.

Samways, M. J., 1981. *Biological Control of Pests and Weeds.* Arnold, London. 58pp.

Sanders, C. J., 1970. The distribution of carpenter ant colonies in the spruce- fir forests of northwestern Ontario. *Ecology 51:* 865–873.

Schemske, D. W., 1978. A coevolved triad: *Costus woodsonii* (Zingiberaceae), its dipteran seed predator and ant mutualists. *Bull. Ecol. Soc. Amer. 59:* 89.

Schemske, D. W., 1980. The evolutionary significance of extrafloral nectar production by *Costus woodsonii* (Zingiberaceae): an experimental analysis of ant protection. *J. Ecol. 68:* 959–967.

Schildknecht, H. & Koob, K., 1971. Myrmicacin, the first insect herbicide. *Angewandte Chemie, New York 10:* 124–125.

Schnell, R., 1967. Contribution a l'etude des genres guyano-amazoniens *Tococa* Aubl. et *Maieta* Aubl. (Melastomacees) et de leurs poches foliares. *Adansonia,* ser. 2, 6: 525–532.

Schoener, T. W., 1971. Theory of feeding strategies. *Ann. Rev. Ecol. Syst., 2:* 369–404.

Schoener, T., 1974. Resource partitioning in ecological communities. *Science 185:* 27–38.

Schremmer, F., 1978a. On the bionomy and morphology of the myrmecophilous larva and pupa of the neotropical butterfly species *Hamearis erostratus* (Lepidoptera, Riodinidae). *Entomol. Ger. 4:* 113–121.

Schremmer, F., 1978b. A neotropical wasp species (Hymenoptera, Vespidae) which guards treehopper larvae (Homoptera, Membracidae) and collects their honeydew. *Entomol. Ger. 4:* 183–186.

Schubart, H. O. R. & Anderson, A. B., 1978. Why don't ants visit flowers? A reply to D. H. Janzen. *Biotropica 10:* 310–311.

Schumacher, A. & Whitford, W. H., 1974. The foraging ecology of 2 species of Chihuahuan Desert (New Mexico, USA) ants, *Formica perpilosa* and *Trachymyrmex smithi neomexicanus* (Hymenoptera, Formicidae). *Insectes Soc. 21:* 317–330.

Schumacher, A. & Whitford, W. G., 1976. Spatial and temporal variation in Chihuahuan desert ant faunas. *Southwest. Nat. 21:* 1–8.

Scott, J. K., 1979. Ants protecting *Banksia* flowers from destructive insects? *West. Aust. Nat. 14:* 152–154.

Scott, J. K., 1981. Extrafloral nectaries in *Alyogyne hakeifolia* (Giord.) Alef. (Malvaceae) and their association with ants. *J. Roy. Soc. West. Aust. 15:* 13–15.

Scott, D. H. & Sargant, E., 1893. On the pitchers of *Dischidia rafflesiana* (Wall.). *Ann. Bot., 7:* 243–269.

Scriber, J. M. & Feeny, P., 1979. Growth of herbivorous caterpillars in relation to feeding specialization and to the growth form of their food plants. *Ecology 60:* 829–850.

Senn, G., 1910. Die Knollen von *Polypodium brunei* Werkle. *Verh. Naturf. Ges., Basel 21:* 115–125.

Sernander, R., 1906. Entwurf einer Monographie der europaischen Myrmekochoren. *Kunglica Svenska Vetenskapakademien Handlingar 41:* 1–410.

Shaffer, D. T. & Whitford, W. G., 1981. Behavioural responses of a predator, the round-tailed horned lizard *Phrynosoma modestum,* and its prey, honey pot ants *Myrmecocystus* spp. *Am. Midl. Nat. 105:* 209–216.

Sharpe, L. A. & Barr, W. F., 1960. Preliminary investigations of harvester ants on southern Idaho rangelands. *J. Range Manage. 13:* 131–134.

Shea, S. R., McCormick, J. & Portlock, C. C., 1979. The effect of fires on regeneration of leguminous species in the northern jarrah (*Eucalyptus marginata* Sm) forest of Western Australia. *Aust. J. Ecol. 4:* 195–205.

Sheata, M. N. & Kaschef, A. H., 1971. Foraging activities of *Messor aegypticus* Emery (Hymenoptera, Formicidae). *Insectes Soc. 18:* 215–226.

Sheppard, C., Martin, P. B. & Mead, F. W., 1979. A planthopper (Homoptera, Cixiidae) associated with red imported fire ant (Hymenoptera, Formicidae) mounds. *J. Ga. Entomol. Soc. 14:* 140–144.

Short, L. L., 1978. Sympatry in woodpeckers of lowland Malayan forest. *Biotropica 10:* 122–133.

Silvertown, J. W., 1980. The evolutionary ecology of mast seeding in trees. *Biol. J. Linn. Soc. 14:* 235–250.

Simmonds, F. J., 1949. Insects attacking *Cordia macrostachya* (Jacq.) Roem. & Schult. in the West Indies. *Physonota alutacea* Boh. (Coleoptera, Cassididae). *Canad. Ent. 81:* 185–199.

Singh, V., Dubey, O. P., Radhakrishnan Nair, C. P. & Pillai, G. B., 1978. Biology and bionomics of insect pests of cinnamon. *J. Plant Crops 6:* 24–27.

Skinner, G. J., 1980a. Territory, trail structure and activity patterns in the wood ant, *Formica rufa* (Hymenoptera, Formicidae), in limestone in northwest England, U.K. *J. Anim. Ecol. 49:* 381–394.

Skinner, G. J., 1980b. The feeding habits of the wood ant, *Formica rufa* (Hymenoptera, Formicidae) in limestone woodland in northwest England, U.K. *J. Anim. Ecol. 49:* 417–434.

Skinner, G. J. & Whittaker, J. B., 1981. An experimental investigation of interrelationships between the wood ant (*Formica rufa*) and some tree-canopy herbivores. *J. Anim. Ecol. 50:* 313–326.

Smallwood, J. & Culver, D. C., 1979. Colony movements of some North American ants. *J. Anim. Ecol. 48:* 373–382.

Smith, C. C., 1975. The coevolution of plants and seed predators. In Gilbert, L. E. & Raven, P. H. (eds.), *Coevolution of Animals and Plants,* 51–77. Univ. Texas Press, Austin.

Smith, J. H. & Atherton, D. O., 1944. Seed harvesting and other ants in the tobacco-growing districts of North Queensland. *Qld. J. Agric. Sci. 1:* 33–61.

Smith, W., 1903. *Macaranga triloba,* a new myrmecophilous plant. *New Phytol., 2:* 79–82.

Sneva, F. A., 1979. The western harvester ants: their density and hill size in relation to herbaceous productivity and big sagebrush cover. *J. Range Manage. 32:* 46–47.

Soholt, L. F., 1973. Consumption of primary production by a population of kangaroo rats *(Dipodomys merriami)* in the Mojave desert. *Ecol. Monogr. 43:* 357–376.

Solbrig, O. T. & Cantino, P. D., 1975. Reproductive adaptations in *Prosopis* (Leguminosae, Mimosoideae). *J. Arnold Arbor., Harv. Univ. 56:* 185–210.

Solbrig, O. T., Cody, M. L., Fuentes, E. R., Glanz, W. & Hunt, J. H., 1977. The origin of the biota. In: Mooney, H. A. (ed.), *Convergent Evolution in California and Chile: Mediterranean Climate Ecosystems; IBP. Synthesis Series 5:* 13–26. Dowden, Hutchinson and Ross, Stroudsberg.

Soulie, J., 1955. Facteurs du mileu agissant sur l'activite des colonnes de recolte chez le fourmis *Cremastogaster scutellaris* (Hymenoptera, Formicoidea). *Insectes soc. 2:* 173–177.

Southwood, T. R. E., 1972. The insect-plant relationship – an evolutionary perspective. In: von Emden, H. F. (ed.), *Insect-Plant Relationships.* Blackwell, London.

Soysa, S. W., 1940. Orchids and ants. *Orchidol. Zeylan 7:* 88–91.

Spanner, L., 1939. Untersuchungen ueber den Waerme und Wasserhaushalt von *Myrmecodia* und *Hydnophytum. Jb. wiss. Bot. 88:* 243–283.

Stager, R., 1931. Ueber die Einwirkung van Duftstoffen und Pflanzen duften auf Ameisen. *Z. wiss. InsektBiol. 26:* 55–65.

Stamp, N. E., 1980. Egg deposition patterns in butterflies: why do some species cluster their eggs rather than deposit them singly? *Am. Nat., 115:* 367–380.

Sterling, W. L., 1978. Fortuitous biological suppression of the boll weevil by the red imported fire ant. *Environ. Entomol. 7:* 564–568.

Sterling, W. L., Jones, D. & Dean, D. A., 1979. Failure of the red imported fire ant, *Solenopsis invicta,* to reduce entomophagous insect and spider abundance in a cotton agroecosystem. *Environ. Entomol. 8:* 976–981.

Stewart, F. M. & Levin, B. R., 1973. Partitioning of resources and the outcome of interspecific competition: a model and some general considerations. *Am. Nat. 107:* 171–198.

Stout, J., 1979. An association of an ant, a mealybug, and an understorey tree from a Costa Rican rain forest. *Biotropica 11:* 309–311.

Stradling, D. J., 1978a. Food and feeding habits of ants. In: Brian, M. V. (ed.), *Production Ecology of Ants and Termites,* 81–106. Cambridge U.P., Cambridge.

Stradling, D. J., 1978b.The influence of size on foraging in the ant *Atta cephalotes* and the effect of some plant defense mechanisms. *J. Anim. Ecol. 47:* 173–1889.

Stringer, C. E., Banks, W. A. & Mitchell, J. A., 1980. Effects of chlorpyrifos and acephate on populations of red imported fire ants, *Solenopsis invicta,* in cultivated fields. *J. Ga. Entomol. Soc. 15:* 413–417.

Su, T. H., Beardsley, J. W. & McEwen, F. L., 1980. AC-217300, a promising new insecticide for use in baits for control of the big-headed ant *Pheidole megacephala* in pineapple. *J. Econ. Entomol. 73:* 755–756.

Sudd, J. H., 1967. *An Introduction to the Behaviour of Ants.* Arnold, London, 200 pp.

Sudd, J. H., 1969. The excavation of soil by ants. *Zeitschrift fur Tierpsychologie 26:* 257–276.

Sunhede, S., 1974. Studies in gasteromycetes. Part 1. Notes on spore liberation and spore dispersal in *Geastrum. Sven. Bot. Tidskr. 68:* 329–343.

Swain, T., 1977. Secondary compounds as protective agents. *Ann. Rev. Plant Physiol. 28:* 479–501.

Takahashi, R., 1951. On the myrmecophyte, *Macaranga triloba,* and its ant in Malaya. *Bull. Biogeogr. Soc. Jap. 15:* 11–12.

Talbot, M., 1943. Response of the ant *Prenolepis imparis* Say to temperature and humidity changes. *Ecology 24:* 345–352.

Taylor, B., 1977. The ant mosaic on cocoa and other tree crops in western Nigeria. *Ecol. Entomol. 2:* 245–255.

Taylor, F., 1977. Foraging behavior of ants: experiments with two species of Myrmecine ants. *Behav. Ecol. Sociobiol. 2:* 147–167.

Taylor, F., 1978. Foraging behaviour of ants: theoretical considerations. *J. theor. Biol. 71:* 541–566.

Tevis, L., 1958a. Germination and growth of ephemerals induced by sprinkling a sandy desert. *Ecology 39:* 681–688.

Tevis, L., 1958b. A population of desert ephermerals germinated by less than one inch of rain. *Ecology 39:* 688–695.

Tevis, L., 1958c. Interrelations between the harvester ant *Veromessor pergandei* (Mayr.) and some desert ephemerals. *Ecology 39:* 695–704.

Teuscher, H., 1956. *Myrmecodia* and *Hydnophytum. Nat. Hort. Mag. 45:* 49–51.

Teuscher, H., 1967. *Dischidia pectenoides. Nat. Hort. Mag. 46:* 36–40.

Theophrastus, C. 300 B.C., *Enquiry into Plants and Minor Works on Odours and Weather Signs,* Vol. 1 (Trans. Sir Arthur Hart). Harvard Univ. Press, Cambridge.

Thiollay, J. M., 1978. Predaceous birds from a zone of contact between savannah and forest in the Ivory Coast; alimentary specifications. *Alauda 46:* 147–170.

Thistleston-Dyer, W. T., 1902. Morphological notes. 7. Evolution of pitchers in *Dischidia rafflesiana. Ann. Bot. 16:* 365–369.

Thomas, M. J., Pillai, K. B. & Nair, N. R., 1980. *Solenopsis geminata* (Formicidae, Hymenoptera) as a predator of the brown plant hopper *Nilaparvata lugens. Agric. Res. J. Kerala 18:* 145.

Thompson, J. N., 1981. Elaiosomes and fleshy fruits: phenology and selection pressures for ant-dispersed seeds. *Am. Nat. 117:* 104–108.

Tilman, D., 1978. Cherries, ants and tent caterpillars: timing of nectar production in relation to susceptibility of caterpillars to ant predation. *Ecology 59:* 686–692.

Townes, H., 1958. Some biological characteristics of the Ichneumonidae (Hymenoptera) in relation to biological control. *J. Econ. Ent. 51:* 650–652.

Trelease, W., 1889. Myrmecophilism. *Psyche, 1889,* 171.

Treub, M., 1883a. Sur les urnes du *Dischidia rafflesiana. Ann. Jard. Bot. Buitenz.* 3: 13–34.

Treub, M., 1883b. Sur le *Myrmecodia echinata* Gaudich. *Ann. Jard. Bot. Buitenz.* 3: 129–159.

Treub, M., 1888. Nouvelles recherches sur le *Myrmecodia* de Java (*Myrmecodia tuberosa* Beccari [non Jack]). *Ann. Jard. Bot. Buitenz* 7: 191–212.

Tryon, H., 1885. Notes on Queensland ants. I. Harvesting ants. *Proc. R. Soc. Queensl. 2:* 146–162.

Ulbricht, E., 1907. Uber europaische Myrmekochoren. *Verhandlungen des Botanischen Vereins der Provinz Brandenburg 40:* 214–241.

Ulbricht, E., 1939. Deutsche Myrmekochoren. *Repertorium novarum specierum regni vegetabilis 67:* 1–56.

Ule, 1901. Ameisengarten in Amazonasgebiet. *Engler. Bot. Jahrb. Syst. Beibl. 68:* 45–52.

Ule, E., 1902. Ameisengarten im Amazonasgebiet. *Bot. Jahrb. 30:* 45–52.

Ule, E., 1905. Wechselbeziehungen zwischen Ameisen und Pflanzen. *Flora 94:* 491–497.

Ule, E., 1906a. Ameisenpflanzen. *Bot. Jahrb. 37:* 335–352.

Ule, E., 1906b. Blumengarten der Ameisen am Amazonenstrome. *Vegetationsbilder 3:* Heft 1.

Ule, E., 1907. Die Pflanzenformationen des Amazonas-Gebietes. Pflanzengeographische Ergebnisse meiner in den Jahren 1900–1903 in Brasilien und Peru unternommen Reisen. *Bot. Jahrb. 40:* 114–172.

Uphof, J. C. Th., 1942. Ecological relations of plants with ants and termites. *Bot. Rev. 8:* 563–598.

Urbani, C. B., 1980. 1st description of fossil gardening ants (Hymenoptera, Formicidae). 1. Attini. *Stuttg. Beitr. Naturkd. Ser. B.* 1–13.

Vandermeer, J. & Boucher, D., 1978. Varieties of mutualistic interaction in population models. *J. Theor. Biol. 74:* 1–10.

Vanderplank, F. L., 1960. The bionomics and ecology of the red tree ant, *Oecophylla* sp. and its relationship to the coconut bug *Pseudotheraptus wayi* Brown (Coreidae). *J. Anim. Ecol. 29:* 15–33.

Van der Pijl, L., 1955. Some remarks on mymecophytes. *Phytomorphology 5:* 190–200.

Van der Pijl, L., 1972. *Principles of Dispersal in Higher Plants.* Springer, New York.

Van Leeuwen, W. M., 1928. Mierenplanten in Mierentuinen. *Nederlandsch-Indisch Natuurwetenschappelijk Congres,* 378–384.

Van Leeuwen, W. M., 1929. Kruze Mitteilung uber Ameisen-Epiphyten aus Java. *Ber. Dtsch. Bot. Ges. 47:* 90–99.

Van Leeuwen, W. M., 1929. Mierenepiphyten. *Trop. Natuur. 18:* 57–131.

Van Leeuwen, W. & Reijnvaan, J., 1913. Beitrage zur Kenntnis der Lebensweise einiger *Dischidia –* arten. *Ann. Jard. Bot. Buitenz. 27:* 65–90.

Vinson, S. B., 1972. Imported fire ant feeding on *Paspalum* seeds. *Ann. Entomol. Soc. Am. 65:* 988.

Vogel, S., 1878. Nectarien und ihre ecologische Bedeutung. *Apidologie 8:* 321–335.

Von Wettstein, R. R., 1889. Pflanzen und Ameisen. *Naturwissenschaft 29:* 309–327.

Vowles, D. M., 1955. The foraging of ants. *Br. J. Anim. Behav. 3:* 1–13.

Wagner, W. H., 1972. *Solenopteris brunei,* a little-known fern epiphyte with dimorphic stems. *Am. Fern. J. 62:* 33–43.

Wali, M. K. & Kannowski, P. B., 1971. Analysis of a saline tall grass prairie ecosystem. Part 3. A study of soil–plant–ant interactions. *Proc. N. Dak. Acad. Sci. 25:* 30.

Waloff, N. & Blackith, R. E., 1962. The growth and distribution of the mounds of *Lasius flavus* (Fabricius) in Silwood Park, Berkshire. *J. Anim. Ecol. 31:* 421–437.

Wallace, C. R., 1967. Seed harvesting ants, *Pheidole. NSW Dep. Agr. Div. Sci. Serv. Entomol. Br. Annu. Rep. 9.*

Wallace, M. M. H., 1975. Insects of grasslands. In: Moore, R. M. (ed.), *Australian Grasslands,* 361–370. ANU Press, Canberra.

Waller, D. A., 1980. Leaf-cutting ants, *Atta texana,* and leaf-riding flies, *Pholeomyia texensis. Ecol. Entomol. 5:* 305–306.

Wallis, D. I., 1962. The relationship between hunger, activity and worker function in an ant colony. *Proc. Zool. Soc. Lond. 139:* 589–605.

Walsh, J. H. T., 1891. On the habits of certain harvesting ants. *Scientific Memoirs by Medical Officers, India, 1891,* 59.

Ward, P., 1965. Feeding ecology of the black-faced dioch *Quelea quelea* in Nigeria. *Ibis 107:* 173–214.

Way, M. J., 1954. Studies of the life history and ecology of the ant *Oecophylla longinoda* Latreille. *Bull. Entomol. Res. 45:* 93–112.

Way, M. J., 1963. Mutualism between ants and honeydew-producing Homoptera. *Ann. Rev. Entomol. 8:* 307–344.

Weaver, N., 1978. Chemical control of interspecific behavior. In: Rockstein, M. (ed.), *Biochemistry of Insects,* 391–418. Academic Press, New York.

Weber, N. A., 1938. The food of the giant toad, *Bufo marinus* (L.) in Trinidad and British Guiana with special reference to the ants. *Ann. Entomol. Soc. Am. 31:* 499–503.

Weber, N. A., 1941. The biology of the fungus-growing ants. Part VII. The Barro Colorado Island, Canal Zone species. *Revta. Ent. Rio de J. 12:* 93–130.

Weber, N. A., 1943. Parabiosis in Neotropical 'ant gardens.' *Ecology 24:* 400–404.

Weber, N. A., 1944. The neotropical coccid-tending ants of the genus *Acropyga* Roger. *Ann. Entomol. Soc. Am. 37:* 89–122.

Weber, N. A., 1947. Lower Orinoco River fungus – growing ants (Hymenoptera: Formicidae, Attini). *Bol. Entomol. Venez. 6:* 143–161.

Weber, N. A., 1956. Symbiosis between fungus-growing ants and their fungus. *1955 Yearbook, American Philosophical Society,* April 1956, 153–157.

Weber, N. A., 1957. Fungus growing ants and their fungi: *Cyphomyrmex costatus. Ecology 38:* 480–494.

Weber, N. A., 1966a. The fungus-growing ants. *Science 153:* 587–604.

Weber, N. A., 1966b. Fungus-growing ants and soil nutrition.

Actas Primero Coloquium Latino Americano Biologico Suelo Monografia 1: 221–256. Centro Cooperativo Ciente Americano Latina, UNESCO, Montevideo, Uruguay.

Weber, N. A., 1968. Tobago island fungus-growing ants (Hymenoptera, Formicidae). *Entomol. News 79:* 141–145.

Weber, N. A., 1969. Ecological relations of three *Atta* species in Panama. *Ecology 50:* 141–147.

Weber, N. A., 1972a. *Gardening Ants. the Attines.* Phil. Soc. Philadelphia, Penn.

Weber, N. A., 1972b. The fungus-culturing behavior of ants. *Am. Zool. 12:* 577–587.

Weber, N. A., 1972c. The attines: fungus-culturing ants. *Am. Sci. 60:* 448–456.

Weiss, F. E., 1908. The dispersal of fruits and seeds by ants. *New Phytol. 7:* 23–28.

Weiss, F. E., 1909. The dispersal of the seeds of the gorse and the broom by ants. *New Phytol. 8:* 81–89.

Welch, R. C., 1978. Changes in the distribution of the nests of *Formica rufa* (Hymenoptera, Formicidae) at Blean Woods National Nature Reserve, Kent England, UK, during the decade following coppicing. *Insectes Soc. 25:* 173–178.

Wellenstein, G., 1952. Sur Ernahrungsbiologie der Roten Waldameise (*Formica rufa* L.) *Z. Pflanzenkr. 59:* 430–451.

Wellington, A. B. & Noble, I. R., 1981. Population dynamics of the mallee, *Eucalyptus incrassata. Abstr. XIII Int. Bot. Congr.,* 268. Sydney, August 1981.

Went, F. W. & Westergaard, M., 1949. Ecology of desert plants. III. Development of plants in the Death Valley National Monument, California. *Ecology 30:* 26–38.

Went, F. W., Wheeler, J. & Wheeler, G. C., 1972. Feeding and digestion in some ants (*Veromessor* and *Manica*). *Bioscience 22:* 82–88.

Werner, F. G., 1973. Foraging activity of the leaf-cutter ant, *Acromyrmex versicolor,* in relation to season, weather and colony condition. *US/IBP Desert Biome Research Memorandum,* RM *73–28.*

Werner, F. G. & Murray, S. L., 1972. Demography and foraging activity of leaf-cutter ants, *Acromyrmex versicolor,* in relation to colony size and location, season, vegetation and temperature. *US/IBP Desert Biome Research Memorandum,* RM *72–33.*

Westhoff, V. & Westhoff-de Joncheere, J. N., 1942. Verspreiding en nestoecologie van de mieren in de Nederlandse bossen. Tijdschr. over Plantenziekten 48: 138–212.

Westoby, M., 1981. How diversified seed germination behavior is selected. *Am. Nat. 118:* 882–885.

Westoby, M., Cousins, J. M. & Grice, A. C., 1982. Rate of decline of some soil seed populations during drought in western New South Wales, this volume.

Westoby, M. & Rice, B., 1981. A note on combining two methods of dispersal-for-distance. *Aust. J. Ecol. 6:* 189–192.

Westoby, M., Rice, B., Shelley, J. M., Haig, D. & Kohen, J. L., 1982. Plants' use of ants for dispersal at the West Head, New South Wales, this volume.

Wheeler, W. M., 1907. The fungus-growing ants of North America. *Bull. Am. Mus. Nat. Hist. 23:* 669–807.

Wheeler, W. M., 1910. *Ants, their Structure, Development and Behavior.* Columbia University Press, New York.

Wheeler, W. M., 1912. Notes on a mistletoe ant. *New York Ent.*

Soc. 20: 130–134.

Wheeler, W. M., 1913. Observations on the Central American Acacia ants. *Trans. Second Ent. Congr. Oxford, 1912,* 2: 109–139.

Wheeler, W. M., 1914. The ants of the Baltic amber. *Schr. Phys. Okon. Ges. Konigsberg 55:* 1–142.

Wheeler, W. M., 1921. A new case of parabiosis and the 'ant gardens' of British Guiana. *Ecology 2:* 89–103.

Wheeler, W. M., 1942. Studies of neotropical ant-plants and their ants. *Bull. Mus. Comp. Zool. Harv. Univ. 90:* 1–263.

Wheeler, W. M., 1973. The fungus growing ants of North America. Dover Publ., New York.

Wheeler, W. M. & Bequaert, J. C., 1929. Amazonian myrmecophytes and their ants. In: *Zoologischen Anzeiger* (Wasmann-Festband), 10–39.

Wheeler, J. W., Olubajo, O., Storm, C. B. & Duffield, R. M., 1981. Anabaseine: venom alkaloid of *Aphaenogaster* ants. *Science 211:* 1051–1052.

Wheeler, J. & Rissing, S. W., 1975a. Natural history of *Veromessor pergandei.* I. The nest. *Pan-Pac. Entomol. 51:* 205–216.

Wheeler, J. & Rissing, S. W., 1975b. Natural history of *Veromessor pergandei.* II. Behavior. *Pan-Pac. Entomol. 51:* 303–314.

Wheeler, G. A. & Wheeler, J., 1978. Mountain ants of Nevada, U.S.A. *Great Basin Nat. 38:* 379–396.

Whiffin, T., 1972. Observations on some upper Amazonian formicarial Melastomataceae. *SIDA 5:* 33–41.

Whigham, D., 1974. An ecological life history study of *Uvularia perfoliata* L. *Am. Midl. Nat. 91:* 343–359.

Whitford, W. G., 1973. Demography and bioenergetics of herbivorous ants in a desert ecosystem as functions of vegetation, soil type, and weather variables. U.S. *International Biological Program. Desert Biome, Research Memorandum RM 73–79.*

Whitford, W. G., 1976. Foraging behavior of Chihuahuan Desert harvester ants. *Am. Midl. Nat. 95:* 455–458.

Whitford, W. G., 1978a. Structure and seasonal activity of Chihuahua Desert, New Mexico, USA ant communities. *Insectes Soc. 25:* 79–88.

Whitford, W. G., 1978b. Foraging in seed-harvester ants *Pogonomyrmex* spp. *Ecology 59:* 185–189.

Whitford, W. G., 1978c. Foraging by seed-harvesting ants. In: Brian, M. V. (ed.), *Production Ecology of Ants and Termites,* 107–110. Cambridge U.P., Cambridge.

Whitford, W. G., Depree, D. J., Hamilton, P. & Ettershank, G., 1981. Foraging ecology of seed harvesting ants, *Pheidole* spp., in a Chihuahuan desert ecosystem, New Mexico, USA. *Am. Midl. Nat. 105:* 159–161.

Whitford, W. G., Depree, E. & Johnson, P., 1980. Foraging ecology of 2 Chihuahuan Desert, New Mexico, USA, ant species, *Novomessor cockerelli* and *Novomessor albisetosus. Insectes Soc. 27:* 148–156.

Whitford, W. G. & Ettershank, G., 1975. Factors affecting foraging activity in Chihuahua desert harvester ants. *Environ. Entomol. 4:* 689–696.

Whitford, W. G. & Gentry, J. B., 1981. Ant communities of southeastern USA longleaf pine (*Pinus palustris*) plantations. *Environ. Entomol. 10:* 183–185.

Whitford, W. G., Johnson, P. & Ramirez, J., 1976. Comparative ecology of the harvester ants *Pogonomyrmex barbatus* (F. Smith) and *Pogonomyrmex rugosus* (Emery). *Insectes Soc.*

23: 117–132.

Whitford, W. G., Kay, C. A. & Schumacher, A. M., 1975. Water loss in Chihuahuan desert ants. *Physiol. Zool.* *48:* 390–397.

Whitten, A. J., 1981. Notes on the ecology of *Myrmecodia tuberosa* Jack on Siberut Island, Indonesia. *Ann. Bot.* *47:* 525–526.

Whitten, A. J., 1982. The role of ants in selection of night trees by gibbons. *Biotropica,* in press.

Wight, J. R. & Nicholls, J. T., 1966. Effects of harvester ants on production of a saltbush community. *J. Range. Manage.* *19:* 68–71.

Willard, J. R. & Crowell, H. H., 1965. Biological activities of the harvester ant *Pogonomyrmex owyheeii* in central Oregon. *J. Econ. Entomol.* *58:* 484–489.

Williams, D. F., Lofgren, G. S., Banks, W. A., Stringer, C. E. & Plumley, J. E., 1980. Laboratory studies with 9-amidino-hydrazones, a promising new class of bait toxicants for control of red imported fire ants *Solenopsis invicta. J. Econ. Entomol.* *73:* 798–802.

Wilson, E. O., 1959. Some ecological characteristics of ants in New Guinea rain forests. *Ecology 40:* 437–447.

Wilson, E. O., 1971. *The Insect Societies.* Belknap, Cambridge, Mass.

Wilson, N. L., Cillier, J. H. & Markin, G. P., 1971. Foraging territories of imported fire ants. *Ann. Entomol. Soc. Am. 64:* 660–665.

Winder, J. A., 1978. The role of nondipterous insects in the pollination of cocoa in Brazil. *Bull. Entomol. Res. 68:* 559–574.

Wing, N. M., 1968. Taxonomic revision of the nearctic genus *Acanthomyops. Cornell Univ. Mem.*

Withers, J. R., 1978. Studies on the status of unburned *Eucalyptus* woodland at Ocean Grove, Victoria, Australia. II. The differential seedling establishment of *Eucalyptus ovata* and *Casuarina littoralis. Aust. J. Bot. 26:* 465–484.

Wood-Baker, C. S., 1977. An interesting plant, aphid and ant association. *Entomol. Mon. Mag. 113:* 205.

Woodell, S. R. J., 1974. Ant-hill vegetation in a Norfolk salt-marsh. *Oecologia 16:* 221–225.

Woyciechowski, M. & Miszta, A., 1976. Spatial and seasonal structure of ant communities in a mountain meadow. *Ekol. Pol. 24:* 577–592.

Wyatt, R., 1980. The impact of nectar-robbing ants on the pollination system of *Asclepias curassavica. Bull. Torrey Bot. Club 197:* 24–28.

Yapp, R. H., 1902. Two Malayan 'myrmecophilous' ferns, *Polypodium (Lecanopteris) carnosum* (Blume), and *Polypodium sinuosum,* Wall. *Ann. Bot. 16:* 185–231.

Yasuno, M., 1963. The study of the ant population in the grassland at Mt. Hakkoda. I – The distribution and nest abundance of ants in the grassland. *Ecol. Rev. Sandai. 16:* 83–91.

Yensen, N., Yensen, E. & Yensen, D., 1980. Intertidal ants from the Gulf of California, Mexico. *Ann. Entomol. Soc. Am. 73:* 266–269.

Yeo, P. F., 1973. Floral allurements for pollinating insects. In: Van Emden, H. F. (ed.), *Insect/Plant Relationships.* Wiley, New York.

Additional references

Ales, D. C., Wiemer, D. F. & Hubbel, S. P., 1981. A natural repellent of leaf-cutter ants. *Proc. Iowa Acad. Sci 88:* 19.

Alvarado, A., Berish, C. W. & Peralta, F., 1981. Leaf-cutter ant *(Atta cephalotes)* influence on the morphology of andepts in Costa Rica. *Soil Sci. Soc. Am. J. 45:* 790–794.

Atsatt, P. R., 1981. Lycaenid butterflies and ants: selection for enemy-free space. *Am. Nat. 118:* 638–654.

Briese, D. T., 1982. Resource partitioning amongst seed-harvesting ants in semi-arid Australia. *Aust. J. Ecol. 7:* in press.

Claassens, A. J. M. & Dickson, C. G. C., 1977. A study of the myrmecophilous behaviour of the immature stages of *Aloeides thyra* (Lep.: Lycaenidae) with special reference to the function of the retractile tubercles and with additional notes on the general biology of the species. *Entomol. Rec. J. Var. 89:* 225–231.

Daniels, G. 1976. The life history of *Hypochrysops theon medocus* (Fruhstorfer) (Lepidoptera Lycaenidae). *J. Aust. Entomol. Soc. 15:* 197–199.

Davidson, D. W. & Norton, S. R., 1980. Ant dispersal of diaspores in some dominant perennials and semi-perennials of the Australian arid zone. *Proc. Int. Congr. Syst. Evol. Biol., 2:* 176. UBC, Vancouver.

Farquharson, C. O., 1921. Five years' observations (1914–1918) on the bionomics of Southern Nigerian insects, chiefly directed to the investigation of lycaenid life histories and to the relation of Lycaenidae, Diptera and other insects to ants. *Trans. Entomol. Soc. Lond.* Parts *III, IV.*

Galle, L., 1980a. Dispersion of high density and nests in sandy-soil grassland ecosystems. *Acta Univ. Szeged. Acta Biol. 26:* 129–136.

Galle, L., 1980b. Niche analysis and competitive strategies of grassland ants. *Acta Univ. Szeged. Acta Biol. 26:* 181–182.

Ito, Y. & Nagamine, M., 1981. Why a cicada, *Mogannia minuta,* became a pest of sugarcane: an hypothesis based on the theory of escape. *Ecol. Entomol. 6:* 273–284.

King, T. J., 1981. Ant-hill vegetation in acidic grasslands in the Gower Peninsula, South Wales, U.K. *New Phytol. 88:* 559–572.

Roepke, W., 1918. Zur Myrmekophile von *Gerydus boisduvali* Moore (Lep., Rhop., Lycaenid). *Tijdschr. Entomol. 61:* 1–16.

Sudd, J. H. & Lodhi, A. Q., 1981. Distribution of foraging workers of the wood ant, *Formica lugubris* (Hymenoptera: Formicidae), and their effect on the numbers and diversity of other Arthropoda. *Biol. Conserv. 20:* 133–146.

Wisdom, W. A. & Whitford, W. G., 1982. The effects of vegetation change on ant communities of arid rangelands. *Environ. Entomol.,* in press.

Wood, T. K., 1977. Role of parent females and attendant ants in maturation of the treehopper, *Entylia bactriana* (Homoptera: Membracidae). *Sociobiology 2:* 257–272.

Wood, T. K., 1979. Sociality in the Membracidae (Homoptera). *Misc. Publ. Entomol. Soc. Am. 11:* 15–21.